THE ONLY GUIDE TO

HVAC SALES...

...that teaches the basics
...worth buying
...that covers it all
...that you don't need a calculator for
...written by someone who's done it
...that you can actually read
...that you can find
...that mentions goats

R J SCHUSTER

Copyright © 2017 R J Schuster
All rights reserved.

ISBN: 1541056183
ISBN 13: 9781541056183
Library of Congress Control Number: 2017902033
CreateSpace Independent Publishing Platform
North Charleston, South Carolina

Introduction

Welcome friends, farm animals, and fellow salesmen of all levels of experience. My name is Rich Schuster, and I'd like to start off by congratulating you for taking this, and frankly, any other step in advancing your career. What's that you say? Your boss is *making* you read this book? Well, curse him, damn boss! Who does he think he is trying to help you raise your income? You know what to do, throw this book in his face and storm out of there screaming like your hair is on fire! Go, now!!! Go!!!

Alright, all those time wasters are gone. I don't need anybody who isn't serious about reading this book, you know. Yes, you're right; this isn't going to be your average, run-of-the-mill HVAC sales book. Then again, I don't think there are that many out there to begin with, sooo…, I guess the comps are going to be pretty good!

I've been around the HVAC (and plumbing) world for somewhere in the neighborhood of 25 plus years now, and one thing that we, (as an industry) never seem to get enough of, is sales training. Now, hold on there, cowboy, I know that most of you *think* that you aren't salesmen, or at least didn't plan on becoming salesmen, but the truth is that most of us are. The sad part is that probably no more than four of you were ever taught the basics of how to properly do your job! Everything we do in this world has a proper technique associated with it as a roadmap to success, and the sales profession is certainly no exception. If you're a technician, you're selling every single day, whether you realize it or not, and if you're a salesman, then, well, duh, and I want to teach you how to do it right.

Let's start with the most basic of the basics; why do we need sales? This answer should be obvious to everyone, but just in case it's not, let's get this out of the way before we go any further. Very simply put, you and your company both need continuing sales to maintain a viable business, and to grow for the future. Everything that you want for yourself or your company, like new trucks, health benefits, office people, job security, retirement, etc., depends on you having work to do tomorrow, and all of it comes from a sale.

This is where someone says, "Not in my company, we just wait for the phone to ring, or for one of our 'fixers' to bring work back." Ok, Pablo, whether you realize it or not, both of those rely on a sale taking place first. The phone only rings because you did some type of advertising and convinced someone that you could take care of their needs (which is a sale), or, one of your "fixers" convinced an existing customer that they needed some extra work that they will now pay for. Either way it's a sale.

Now that first sale has more to do with marketing, which is really a science in itself. Plenty of people have made a career out of trying to figure out just what it takes to make someone pick up the phone and call you because of the words and pictures that you put in front of them. We'll get into a little bit of that in a later chapter as part of our discussion on advertising.

However, our friend Pablo also said that his fancy technician might also bring back a work order from an existing customer. Now that very much has to do with what we're going to talk about here. That is a one-on-one sales encounter that is done millions of times a day, and quite literally makes our economy work. Despite the fact that Pablo's boy probably doesn't even know what happened, his sale will provide the fuel that helps your company live another day.

Just because he isn't wearing a white shirt and hasn't combed his hair in a week, doesn't mean that he isn't a salesman; he most certainly is, and so are you. We all are. Every day that you're alive, you are trying to convince someone to do something that wasn't their idea. It might be something as simple as convincing your buddy to try the new burger joint across the street, even though he's a vegan. Or maybe one of you are still trying to get your mom to let you stay up and watch Conan (you know who you are). But, for now, we're going to stick to the idea of increasing our HVAC sales as actual paid salesmen.

Let's get right in to it. How many ways do you think there are to increase your sales, in any sales category; condensers, filters, beer, girdles, 50mm cannons, chickens? What do you think; a hundred, a thousand, a gazillion? Nope, just two. That's right; there are only two measly ways to increase your sales. You can either sell more stuff to the customers that you already have, or you can find more customers to sell your stuff to. Now that I've said it, you're saying to yourself "of course, I knew that". Well, you're right, I'm sure that you did know that. You know, none of this is magic, as I like to say.

Most of what you're going to read in this book is either pretty simple, or stuff you've heard before, but just never put it all together in one big coherent thought; which brings me to my second favorite saying: "this is not rocket science". It really isn't; think about it. You know plenty of salesmen making a decent, or actually, a very good living, and half of those guys are potato heads! You know I'm right. Look at that guy over there with his stupid hair and fancy pen. And what does he do all day? Nuthin'. Well, that's not exactly right, he did figure out how to close a sale well enough to make a living. But, if *that* guy can do it, you should be able to kill it!

Before you can make a sale though, you have to have a sales lead of some kind. A sales lead is simply information about a person or company that has a need for, or has expressed interest in, some product or service that you offer or sell. Sales leads can come from any number of sources, but not always easily. Companies spend huge amounts of money trying to generate leads and appointments for their sales force, which makes it critically important to get as much out of every single lead as is possible.

Most of your competition are guys who were technicians for a larger company, eventually got tired of working for someone else and decided to start their own business. Lucky for you, most of them never even considered any type of sales training, which automatically gives you a leg up. While many of the steps in the sales process may be obvious to some, putting together an effective sales program in a company is time consuming and not what most new self-proprietors consider important. Unfortunately, without some strong guidance and dedication, these young companies will likely be subject to years of inconsistent financial success and untold frustrations. A good sales program could change their lives, but they won't see it yet. Oh well, more for you.

Unlike those guys, the days of sitting in the office and waiting for the phone to ring, for you, are gone. Today's businesses have too many expenses and obligations not to grab fate by the shorties and say "ten percent of that business out there is mine, and I'm coming for it!"

That's why you're here; to learn how to put together and execute an effective and efficient blueprint of how to increase your company's sales. Everything from putting a plan together, advertising, collecting leads and qualifying them, to actually running the appointments, fact-finding and doing an estimate, or proposing a solution. Then, of course, you have to close the deal.

But don't stop there; you're far from done if you want to be good at this. You now need to help them figure out how they'll pay for the work, and then you need to make sure that they get what they paid for. I don't care if you have a project manager to handle that; whose customer is it, anyway? It's yours and you MUST stay involved until the project is done, and even after. And, as you'll find out, you're still not done. Don't think of ever being done. Part of the secret to all of this is to never being done. I want you to build relationships that will last for years, and make your job easier the longer you're in it.

But, as you'll see that I often do, I'm getting a little bit ahead of myself here. You need to be aware of, and learn how to manage not just the nuts and bolts numbers, forms, and processes, but also the intangibles of the sale. The intangibles are things like, do they like and trust you, do they believe that you know what you're talking about, and just as importantly, what are they *afraid* of. Yes, they are afraid! And the more of these pieces that you can control, the better your closing ratio will be. If you just said "closing ratio, what's that", then you have an awful lot to do from now until the end of this book, so stop playin' Candy Crush, or working on your fantasy football league, put down the cookie and get to work!

What's the Plan (Goal)?

Here comes the (first?) worn out cliché "If you don't know where you're going, you'll never get there". Yuchhhh. Well, it is true, you know. You could have all of the talent, skills, knowledge, and resources in the world, but if you don't have a plan, you're guaranteed to, at least, waste an awful lot of time and money getting there, if you actually get there at all. Let's just cut the crap; you need a plan, and you need it now. Long term, short term, you need to figure out what and how much you want to sell, and how you are going to do it, regardless of what position you play in your company.

There are actually many plans that you should have in place in your life, depending on how structured you are and how successful you want to be. Or, I guess you could just wander around aimlessly for a while, like a, a, uh, I don't know, like a duck. (shrug). Well, anyway, plans like, where you want to live, whether or not you want to get married, how many kids you want to fund (raise), if any, what to do in retirement, where to put the umbrella stand, etc, are just a few. But let's leave those things up to you and your shrink, and focus on business.

This may be a subject (plans and goals, that is) that you've never given much thought to, but is now suddenly and completely overwhelming. Where do you start? Which goals are most important? What do you do if you "don't have a PhD?" you ask. Well, relax; like everything in this world, we're going to take it one step at a time, and none of us have PhDs.

First of all, your personal goals as a salesman or technician are going to be different than the company's goals, but they are intertwined. So, before you can work on yours, you should look at where the company wants to go. Things won't work out so well if you want to sell ductless mini-splits all day long, but the company wants to specialize in scotch-marine boiler repair, right? So, you need to be sure that you're on the same page first. The problem with that statement is that there is a good chance that your company isn't much more organized than you are. If that's the case, you should get together with your manager or the owner, and tell them what you're trying to do, and ask them about their vision for the company's future. Maybe you can help shape the goals for the company. I'll bet you that a lot more comes out of that conversation than you expect.

Let's assume that there is some sort of vision lurking in management's psyche. Imagine that when you ask Mr. Chromedome where he sees the company going, he suddenly gets that sparkle in his eye and tells you that he just landed a big contract, and that he's splitting the company into two. Now he must buy a larger building, and therefore he expects you to increase your sales. Oh, by the way, he's hired another installation crew, a full time marketing person, and doubled the advertising budget. And, in addition, "we're going to start job-costing all of our work, because we need to figure out how to net another 5% on all of our installs". Surprise!

Or, what if he puts his head in his hands and says that he's hearing that new regulations are coming, and installing water heaters are going to be more trouble than they're worth, so they're going to stop selling them, come March 1st, as a tear gently rolls down his cheek. Never mind the fact that water heaters comprised 30% of your sales! Yikes! Now you can begin to see how a little context may help shape your plans.

Beyond any revelations that you may uncover during this meeting of the minds, you will likely find that your supervisor is pleasantly surprised at your questioning, and very happy to learn that you are taking a proactive step in your future, and the future of the company. Just like that, you have lit the candle in his head that suggests that you might be a good sales manager one day, too. See, all you did so far is ask one little question, and already you're up for a promotion!

I know that I was sort of making a joke there, but only "sort-of". I want to take a quick time-out to emphasize the importance of speaking to your manager/supervisor and letting them know what you're thinking. Way too many times in my career I've heard employees talk about how frustrated they were because "the other guy got promoted instead of me". Go look in the mirror, because that's your fault. I can say with 99% certainty that they chose the other guy for one or more of three reasons. One, they had a better attitude, two, a better work ethic, or, three, as in this case, showed *and expressed* better interest and vision for the company's future than you. *You*, not them, control your future. Take responsibility for your life and your results.

Ok, enough preaching for the moment (yes, there will be more), let's get back to the goals and plans. So, to surmise the last few paragraphs, you need a goal to shoot for, but not just a random goal, you need a well thought out, achievable goal that is acutely aligned with the goals of your company. Basically, you need to make sure that you're all pulling on the same end of the rope. (No, there's not *actually* a rope, it's just an expression! Really, have you ever seen a "sales goal" rope? Come on, now!)

Further, there exists, in the universe of goal and plan setting, a few cardinal rules that must not be broken. These are not at all secrets; in fact you will find that virtually all books, programs, teachers, etc. espouse the following three components. The first of which states that the goals *must (read MUST!!!)* be specific and obtainable; any questions? Yes, you do; don't lie to me. In setting a goal in the sales world, being specific is easy; it means how much or how many of something. For instance, how many dollars in sales do you want this month or year? Or, perhaps, you may want to sell a certain number of units in the first quarter. What's that, you don't sell units? Well substitute the words "UV light" in there for now, Skippy. And the number has to be realistic. Don't tell me that you're going to sell 800 units, err, excuse me, 800 "UV lights" this month if you've never sold more than three in a four week span. How about shooting for ten this time? The point is, it must be clear and measurable (specific), as in ten pieces in a four week period, AND it must be obtainable; ten seems to be reasonable in this case.

The second cardinal rule of undeniability (?) is that you *must (read, MUST!!!)* review your progress often. This may be a daily activity, weekly, or whatever makes sense to you, but it must be done regularly. And be honest with yourself here. I'll let you in on a secret; you're going to get it wrong. There's no shame in that, a lot of what you're going to do here is nothing more than an educated guess, but that's ok. We all do this. Goal setting is an inexact science, and it can't be anything more. It's a destination; we're just not exactly sure how or when we're going to get there. So, review often.

The third and final rule is also a simple one; that is, adjust as necessary. That's it; after you review your progress, just figure out what's working and what isn't, and change some stuff. How's that for technical, "change some stuff". Well, that's what you need to do. That may mean adjusting your tactics, or revising the sales dollar amount or number of units (!), or maybe even the time frame. Maybe you could try a different approach, or a different closing technique, or a different price, do something different. Hell, try a new cologne for Gumby's sake, or maybe comb your hair this time, or try not being drunk when you go to the next sales call; who knows, but keep track of what you learn.

There are many, many areas that you can set goals for in your business world, and the more that you think about it, the more you'll see what I mean. You could set a goal for sales dollars, gross profit %, number of units, how many new customers you get, how many cold calls you make, how high you can get your closing ratio, how many service calls you can add-on to, how many equipment leads you can bring back per day, etc, etc. There is no end to what you can do here, and I would encourage you to take it one step farther and create a competition with a fellow salesman, or how about making a bet with your boss? Bet him that if you can sell four media filters this week, he has to come to work wearing his Sponge Bob pajamas for a week. I don't know, it's your bet, don't ask me.

OK, now you've got a goal, or 62 of them. Now what? Well, having a goal is a great thing, in fact, if you really want to do some heavy duty research, you'll find a whole slew, yes, slew, of books on goal setting. I would go so far as to say that it may actually be an entire industry. But the interesting thing that I find is that the experts will tell you that once your goals are set, and you *write them down (this part is important)*, your subconscious will actually help you obtain those goals. It may do any number of things, like help motivate you to make another phone call, or try closing one more time, or maybe even use a different expression when speaking to a client, but the point is that this part really *is* like magic. I don't know the science behind it, and I frankly don't care, but the experts agree on this phenomenon. I always say that "it's not magic", but in this case it really is!

Once again, now you have these goals, what do you do next? The next step is, as I alluded to above, to plan out exactly *how* you are going to go about achieving these goals. In other words, "What's the plan, Stan?" A goal is just a dream if you don't have a comprehensive plan on how you are going to get there. If you're in Charlotte, North Carolina and decide you want to go to Las Vegas, you get on your bicycle (you lost your license in the flood), and start riding west without a map or directions, what are the chances that you'll actually get there? (No, you can't ask for directions because you're a deaf mute, and you can't write it down because you lost all of your fingers wrestling that alligator in Louisiana, and, just for good measure you're blind in your left eye, and ya got pink eye in the other. Face it, you're a mess). Your chances of getting to the Bellagio are about as good as the Browns getting to the Superbowl in your lifetime; zeee-ro.

What you need, my friend, is an actual plan. Ah, yes, the plan will set you free. With your goals written down, and in your pocket, you can now figure out a plan of how to get there; a road map, if you will. A plan is the difference between a pipe dream and a goal. Every other teenager says he's going to retire by the time he's thirty five, but none of them have a plan, or even a clue. But, you know, every once in a while you find that kid who keeps to himself, has a goal, and a plan. He keeps his plan in his pocket, and never forgets where he's going, or how he's getting there. That's the one who makes it. He's the guy in the news, or on the commercials; the man with the plan, Stan.

Back to the real world for now, there's no need to be intimidated by the thought of making a plan to reach your goals. Keep it simple. If you have kept track of your sales results in the past, you'll be able to see that it's all in the numbers. If you take the number of appointments you ran over the course of the last year, let's say, and divide that by the number of sales you've closed, you'll probably discover that your closing ratio is about 30-35%, or one sale for every three appointments run. That makes it easy, now you know how many appointments you need to run to get X number of sales.

By the way, guys are always trying to tell me that they've got a "75%" or "90%" closing ratio; they don't. Unless their leads are already sold, or the customer calls in and says that they're ready to buy, and the salesman just has to get the work order signed, the average salesman is going to have a 30% to 35% closing ratio, and a great one may be around 40%. Any higher than that, they're either cherry picking the calls they go on, or their prices are too low, and they're leaving money on the table. You know the company that I'm talking about, they're always really busy, but they go bankrupt every two years.

So, go back two paragraphs. If you want 10 sales, then you have to run 30 appointments. If you want 150 sales, then go see 450 people. It IS that simple; go talk to a statistics major. The numbers don't lie. Of course, you have to start with a large enough number to make the averages work. Don't go and see three people, then complain to me that you didn't close any of them. Let's talk after you run the 30 leads.

Now, if we take this idea to the next step, you can really hone in on some exciting data. If you go back to your sales history and look at the last one hundred sales that you made, take out the three biggest and the three smallest, add them all up, then divide by how ever many you have, you will be left with an average dollar-per-sale number. Take that number and divide it by your closing ratio, and you will be left with essentially how much each appointment run is worth to you in sales dollars. That sounded like a lot of mumbo-jumbo, so let's do it again. If your average sale is $6,000.00, and your closing ratio is 33%, then each appointment run is worth the equivalent of $2,000.00. Then, by extension, if you run 50 appointments this month, you should have about $100,000 in sales. That's the way it works; period.

> **Closing Ratio** = number of sales closed (sold) divided by number of appointments run
> **Average Sale** = total sales dollars divided by number of sales (first remove the two or three highest and lowest)
> **Each Appointment is Worth** = average sale multiplied by closing ratio

In my very early years, I worked for a large life insurance company as an agent, which turned out to be three of the most valuable years of my sales life. Just imagine an actuarial company that is involved so intensely in sales. They had statistics on everything! All of the new agents would be required to come in for three hours, two nights a week to be telemarketers for themselves. We would have a list of names and phone numbers to call and give our pitch to. We weren't selling a product, or insurance, we were just trying to set appointments to get in front of people so that we could then try to sell our products. So, in other words, the first sale was the appointment; remember that, there will be a quiz.

This seemed like such a trivial, and random task at the time, but as I would learn, it wasn't random at all; quite the contrary. You see, for every 100 people that I called on any given night, about thirty people would answer. Out of those thirty people, I was able to set about three appointments. When it was time to run those three appointments, one would cancel, and I would close one of the remaining two. Now, of course it didn't always work out exactly like that, but that's how the averages worked out. Then to take it a step further, my average commission was about $600 per sale at the time. So let's do the math. If I closed one out of three, and my average commission was $600, then each appointment that I set was worth about $200. I had to speak to ten people to get one appointment (before I said I had to speak to thirty people to get three appointments, that's the same thing), so by doing the math, we could say that each person that I *spoke* to (not called) on those telemarketing nights was worth about $20.00! And if you want to go back even further, I said that I had to dial one hundred people to speak to those thirty (30 divided by 100 = 30%, then 30% multiplied by $20.00 = $6.00), that means that every time I dialed the phone on those lonely nights in the office, I was rewarded with $6.00! This may be more math than you care to do, but this is real, and you can figure out what each and every appointment or phone call is worth to you too; just keep good records.

Plain and simple, that is a road map. It's not, by any means, the only one you could use, but that is an example of how you can start your plan. The important thing is to try and put together a plan that is based on some sort of history; something that you know to be true, and build on it. Maybe you'll want to tie it into some advertising that you know works for you, like leads from a home show, or an air conditioning ad that you run on Facebook. All you need to do is track the results, and you've got the beginning of a plan.

Let's go in a little different direction for a few minutes now. You have your goal(s), and the beginning of your plan on how to achieve these goals. We've generically discussed or referred to your customers, but who exactly are these so called "customers"? It's a big, wide world out there with lots of buildings of all types that all need work; which do you, or will you focus on? Maybe you've focused on residential work up to this point, but have you considered other types of work? I'm not going to tell you that you should or shouldn't, but I am going to tell you that you should at least get more information so that you can make an informed decision.

One thing that makes sense here is to figure out what you're good at and do more of that. Believe it or not, we're all good at something, meaning each of us as individuals, but also as companies. Some companies excel at selling and performing service, while others would be better at breeding mountain goats than doing a service call. The same goes for installs; some companies are great at installing central air conditioning systems in a basic three bedroom ranch. They can knock them out in one day with three men, do it right, clean up, and make 50% GP all day long. Others are better at the custom six bedroom, 5500 square foot home with four systems, with all the bells and whistles. While others still, kill it on ductless mini-splits, or $100K geothermal systems. It doesn't matter which box you live in, as long as you can see the label on the box.

What about the non-residential work? There are all kinds of commercial, industrial, institutional, and even government projects out there for the taking. But each of these requires a different sales skill set, or different procedure, at least. If you've only worked in the residential world 'til now, you may be asking "why should I bother", and that's a reasonable question. Well, let's look back at your work load over the last two years. There have been a few periods where you had so much work that you could have used four more guys, that's true. But there have been at least as many weeks when you sat around the office, getting uncomfortably nervous, because it was so slow, you started to wonder if you were going to be able to make payroll again. No fun.

By spreading your wings a little, you can smooth out those bumps in your commission and your work load. These other areas of work may be quite different than you are used to, though. Larger, commercial work takes a long time to develop, but can be very rewarding, once you do, partly because you have much less competition. This type of work has a different pricing structure that you'll need to learn, and most of that will come from good old trial and error. To start with, your margins will be lower than you're used to, but that's not necessarily bad. Often times you can work on these jobs in between others, when you would normally have nothing to do, so in the end, you're sort of filling in the blanks.

Secondly, many of these jobs have built in add-ons, or additional work that you will be awarded, just because you are there, without regard for the price; that means you win! But unlike residential estimates, these often take months before the work is awarded, so, if you choose to give it a try, put a few irons in the fire (quote a few different jobs).

Then there is government and institutional work. This could be some type of service work, like being responsible for doing a tune-up on all of the rooftop units on a college campus. This type of work is usually put out to bid months in advance for the next calendar year or two. Once again, it may be worth doing the tune-ups at a very discounted rate, just so that you can get all of the repair work. How many rooftop units are on the college campus down the road from you; seventy five? Don't laugh; it's not out of the question. Again, with this type of work in your pocket, your guys will always be busy, and that sounds like the jingle of commission dollars, my friend.

Ok, so now you've put together a goal that includes how much you want to sell this year, broken down by month. You've gone back through the company's records and figured out what percentage of those sales should be attributed to each month (seasonality, you know). Then, you developed a plan and decided that you're going to try your hand at some of the other types of work out there and maybe bid some of them fancy government contracts. Don't stop there, talk about how much you want in "add-on" revenue each month. By that I mean, humidifiers, media filters, UV lights (yes, ok, units), electrical upgrades, or whatever else you guys do, put that stuff in your goals and plans! Put a target on your closing ratio, and while you're at it, challenge that guy sitting next to you. Tell him that you bet him $100 that you can sell more thermostats than him this month; watch what happens, you'll both sell twice as many as you normally do.

One more thing that you should do, that is to keep score, and make it public. This is all part of what I talked about earlier when I said that you need to review your progress often. Put a score board up in your office, literally. Get a big, stupid 6'X6' dry erase board with each salesman's name on it, and list all of their sales each day. And, make sure that you have their goals on there too; all of them; sales goal, number of referrals, GP%, whatever. Do that, and every single category will improve, and the guys that don't improve will probably quit. That's ok; more for you.

<u>Take-Aways</u>:
- Set specific goals, write them down
- Develop a detailed plan (road map)
- Review them often

- Adjust as necessary
- Diversify the types of work you do
- Keep score

Professionalism – be honest, organized, neat and clean, prompt, and accessable

Be aggressive in your pursuit, not at the table.

Advertising/Marketing; Mining for Gold

Congratulations, you've set a goal and made a plan, or at least hopefully now see the need to do so. That's an important step in your success, so take it seriously and dedicate an appropriate amount of time to do it right, but let's continue. From here, we're going to talk about the mechanics of the sales process, no power tools needed.

As with most activities in life that are purposely repeatable, if you can identify and repeat a specific series of events, you greatly increase the likelihood of your desired outcome. Then, also, you create the opportunity to increase the efficiency of said events to again better improve the desired outcome......Err, read that again, that might not make sense, I just woke up. Uh, let's see, ahhh, blah blah blah, blah blah blah. Nope, it's fine. Or, another way of saying it is, "If you do the same thing over and over again, you can get better at it".

Conveniently, "sales" happens to fall into the category of a repeatable series of events. If you deconstruct virtually all sales of any kind, you will discover that most follow a specific process, even if it's not always obvious. In manufacturing, you may do some research, make the product, advertise it, fill an order, and collect the money; simple. If you sell automobiles, first you do some advertising, maybe then you put on a gorilla costume wearing a bathing suit and do a TV commercial and talk about how your prices are "bananas!" Then, customers come in, you talk them into buying stuff they don't want to buy, have them sign a bunch of unreadable paperwork, and get them financed for 132 months at 14%, and your sale is complete.

In the contracting world there are a few more steps, but the costume is usually optional. After deciding what you want to sell, you do some advertising to get your phone to ring. You set up an appointment with your prospective buyer to uncover their wants, needs, and problems, then propose a number of possible solutions in the form of an estimate. Next, you close the deal and negotiate the terms of payment, complete the work, and get paid. If you plan on doing this for more than a week, you'll want to include multiple follow-up contacts, too.

First on our list of action, you'll notice, is advertising; that is assuming that you already know that you're an HVAC contractor and what it is that you want to sell. Advertising is a wide ranging and ever evolving subject not to be taken for granted. Gone are the days, though, of just placing an ad in the newspaper and phonebook (that was a big yellow book that had everyone's phone number listed, and was also used as a booster seat for your dad when he was a youngun') and waiting for the phone to ring.

Some of you may be asking yourselves "why do we need to get involved with advertising anyway?" Well, if you've gotten this far without any sort of marketing, good for you; but you've been lucky. There are times in business that you have more work than you can handle, and that's great. But there are also those times when you wonder to yourself if business will *ever* return to normal, and how you are going to eat this week. I'm serious, this business is one big rollercoaster, and anything you can do to smooth out the waves will help you live longer, and a good advertising program is one of those things.

Advertising is much more than simply letting the world know that "Little Joe's HVAC and Raccoon Removal" is out there. No, most advertising can be way too expensive to just throw something together haphazardly and hope for the best. First, you should ask yourself what it is that you're trying to accomplish, or *what the goal is* (yes, again). Don't hurt yourself, now, I'm going to tell you. Your advertising goal is to first make the world understand that you exist, yes, but secondly and more importantly, is to get them to pick up the phone and call *you* to solve their problem; *to take action*. Advertising = leads; leads = sales, sales = money, money = _____(go ahead, fill it in)! A good source of leads is like owning a gold mine.

If you were a manufacturer, or something other than a contractor, you may only be interested in brand awareness. There are many types of businesses that would never expect to have the intended audience make contact, or take action when seeing their ad, such as, I don't know, Singapore Airlines for instance. I don't think they really expect someone who is lying on their couch watching Bob's Burgers to suddenly jump up screaming "quick, honey bring me the phone, we're going to Thailand!" because they saw a TV commercial. Of course not; their goal is to make sure that you remember "Singapore Airlines", and how luxurious their service is the next time you plan a trip. Just like ketchup; nobody says "hey, I just saw a ketchup commercial, let's go get some!" No, Heinz is trying to get you to recognize their brand on the shelf the next time you go shopping.

As your company grows in size, brand awareness advertising may be something that you specifically pursue in an attempt to drive business to your service department ("We're ready 24 hours a day!"). But for this conversation, you need your potential customers to take action and call you, NOW! You need leads that you can follow up on and turn into sales NOW! And, you need a lot of them. So, when you begin to put your ad(s) together, take your time and think about the goal.

What should your ad look like? Now, I want to be clear that I am not a marketing expert, but I have been involved in this for a long time and have learned a thing or two about what works and what doesn't. And one of the first things that I learned is that nobody wants to take the time to read your ad; they just don't, they have better things to do. With that in mind, you need to create a simple enough ad so that your potential customer can get the point that you are trying to get across as they are turning the page or throwing your direct mail piece in the garbage. They aren't going stop to read the three paragraphs you wrote about how great your company is or that you only use MERV 13 filters instead of 1Is. In fact, I would tell you to limit the number of words you use to somewhere around twenty.

Simple works; say things like "Need a new A/C system? Call 555-1234 NOW and save!" or "Call now, and be cool by the weekend!" Too often we get wrapped up in industry jargon, or that "we only use "cryronaflowbeneze technology in our smegleflux installations". Most won't know what you're talking about, and they probably don't care. They just want to be cool.

There are many key words and phrases that can be used to attract attention and cause someone to take action. As always, this isn't magic, I'm talking about words like "SALE" and "DISCOUNT", or "FREE" and "PACKAGE". "Package" is an interesting one. People love "packages" and "bundles", like an "Energy Efficiency Package" or the "Indoor Air Quality Bundle" that you may want to focus on in the springtime. You don't want to explain everything in your ad, in fact, quite the opposite, you just want to give them enough information to make them want to call and find out more. Remember, the purpose of your ad is to get people to call in and set appointments.

Another thing to consider when designing an ad is relevancy. By that I mean, is everything that you want to put in your ad really necessary? I know that your 1952 Studebaker is the pride of your life, but do you really think that is going to help someone make the decision to call you for air conditioning? No, don't be stupid, they don't care; get rid of it. And that goes for the goat, too. Go put that picture back on the wall in your bathroom, and stay focused on the goal. Keep your ego out of it.

Before we finish talking about the ad itself, consider other methods to inspire action, such as a deadline. A sale price or benefit of some kind that expires at a certain time or date will create a sense of "I better do it soon, or I'll miss out" mentality. Most people, even if they aren't interested in what you have to offer, will feel a momentary jolt of "I'd better hurry" syndrome. A simple "10% off until the end of the month", or "Free media filter (a $500 value!) for the first 20 customers" may double your response.

Ok, you've got your ad, sans goats and all, where are you going to put it? Well, as I said earlier, "times, they-be-a-changin'." The phone book is dead money, don't even bother trying it, and newspapers ads are questionable. But first, you should set a budget and have a plan of which ads you want to run and when. You may have a few different products or versions of an ad that you can rotate or change with the seasons. (You're just becoming a planning fool, aren't you!) The key here is consistency. Running an ad for a weekend somewhere may get you a few calls, but if you run the same ad consistently in the same place for say, two to four months, your cumulative results should improve. People often need to see the same ad a number of times before they internalize it and take action.

Setting an ad budget for the year, spending $500 per month, for instance, will help keep you on track for consistency. When you decide where your ads will run, committing to a greater number of months or ads will usually get you a good sized discount over just a single ad. The advertising department of wherever your ad will run will be able to guide you for the best possible results, so make sure that you ask.

But *where* should you spend all of this advertising money? Well, that's a good question. More and more, the world is focused on-line. Every day I'm amazed that some web site on my phone is showing me an ad for a local business in my neighborhood. I think that venues like Facebook and Google are absolute necessities for any business these days, in conjunction with a good website. On this subject in particular, I'm going to be vague because it's all changing to quickly, and anything that I say could be out of date by the time this is printed. So, let's just say that you need to stay with the times, and be very aware of what is happening with technology for the majority of your advertising dollars. Take that as a directive to focus heavily on social media, this is not an option.

This is not to say that all of the traditional methods are useless, not at all. I love billboards, for example, and local television and radio work well. In addition, direct mail can still be affective, as well as the Val-Pak type advertising. When you begin working on your advertising campaign, consider speaking to the representatives of the manufactures' products that you are selling about the availability of co-op funds. These are funds that are set aside by the manufacturers to help you with your advertising costs if you mention them in your ad. In addition, they may have complete ads ready to go, and they will put your company's name and logo in for free. This may or may not be what you want to do in the end, but you should at least find out what is available.

<u>**Take Aways:**</u>
- Have a plan and a budget
- The goal of advertising is to get them to call you; to take action NOW!
- Leave the "brand awareness" advertising for the service department
- Focus on social media
- Design your direct mail ads for them to see and understand as they are throwing it in the garbage
- Resist the temptation to add a picture of your goat

Leads – Your Stash of Gold

Without a lead, you have no place to go. You're like a firetruck without a blaze, a doctor without a patient, a politician without a press conference. You're just a lost piece of toast; you poor, moldy bastard. Yes, leads are the air that salesmen breathe. Lucky for you, you've decided to climb the ladder of life and buy this book! Alright, enough applause.

OK, let's get to the gold part. You did some advertising and marketing, gave someone a business card, and wrote an ad on the back of a paper plate and hung it on a telephone pole. The leads are coming in now at a blistering pace. Now what? What are you going to do with them? How will you keep track of them? With that new ad on the telephone pole, you've grown out of the shoe box. What to do, what to do, what to do?

Before we get ahead of ourselves, let's actually define what a lead is. I think this is important because you can make numbers, graphs, and charts do almost anything you want, depending on the definitions that you apply to the data. (I used to know someone that loved the word "data"; isn't that weird? How can you love a word?) Facts and numbers can be squishy things, as you'll see next.

Have you ever come across a guy who likes telling people that his closing ratio is, like, seventy five percent? And he believes it. Well, it could be true, sort of, but he isn't giving you all of the facts. He may be closing seventy five percent of the calls he goes on, but what type of calls is he being given? Is he being given only the leads that have called the office saying "Hi, one of your technicians gave me an estimate to change the batteries in my thermostat last year, and I'm ready to buy now." Or, how about this one "Hi, yeah, uh, lightning just blew up my condenser, and uh, we're having a wedding here this afternoon for my cousin's nephew, you know. Can you send someone to change it out right now? No, I don't care what brand it is, and cost doesn't matter, either; oh, and I'm paying cash". Now Mr. Super technician looks like an imbecile because he can *only* close seventy five percent of his leads. Don't buy into his bull crap that easily.

So, what is a lead? Well, a lead to me is the name and phone number of someone that has expressed interest in getting more information on a product or a service that you offer or sell. Most of the time these leads come from a response to some type of advertising that you or perhaps, a manufacturer has done. This could be from a direct mail piece, a radio ad, a placemat in the diner, a referral from a friend, or any number of places. It is *not* someone acting on a previously given estimate, or someone who has finally figured out how to pay for a new filter, and just needs someone to get the signature.

Leads can be broken down into categories even, if you like. For some reason we relate them to temperature (no, it's not because of our industry) such as in cold, warm, and hot. A cold lead would be someone that doesn't have a strong need to buy now, or doesn't know you from a hole in the wall. They may or may not buy soon, who knows. These are people who think that they may want to add central air conditioning to their home someday, or who know that their furnace or boiler has seen better days, and that they should probably start thinking about replacing it soon; or they might go to Aruba.

Cold leads are the most challenging because the potential buyer has no strong sense of urgency. This type of lead should comprise most of a salesman's work load. Sure, we all would love to just get the "lay downs", but our bread and butter is working and developing these people from ponderers into purchasers. This is why you have a job in the first place. Your boss doesn't need to pay a salesman to just go around and pick up signatures and checks; he's got his bone-head son-in-law to do that. He needs someone with mad sales skills to turn those dreamers into new equipment owners! He's counting on you!

Next, you have warm leads. These can again come from almost anywhere, but there is an increased sense of purpose here. These people are sleeping in a hot room now, or not sleeping at all, for that matter. They've been repairing the same heater again and again, and are tired of it breaking down, or their fuel bill came in a box this time and they want a more efficient unit. They may have done some research, and are close to buying. All of the people in this category are close to buying, and are waiting for someone like you to make them feel like it's the right decision. These leads are coveted by salesmen, and well they should be. These could be second or third appointments for you; places where you've laid the ground work, and now comes the payoff.

Warm leads could also be considered to be appointments with friends or family, or people somehow connected to your company. They trust you for some reason, and will likely buy from you when they are ready. This is a non-hostile environment, unlike those others who won't even let you in their house. You know what I'm talking about; the guy who wants to talk outside in the rainstorm, and he's got an umbrella and you don't? No, these people let you in and offer you a cup of coffee.

So, what about hot leads, you ask. Hot leads are the ones that make you call your wife and say "Honey, put the Spam away, and put on yer fancy sneakers, I'm takin' you to McDonalds tonight! Yeeehaaww!!!" Isn't that what you say? Anyway, hot leads are people that need to buy right now. Something is broken or happening that makes it imperative that they act today. These are the ones that make someone think that they have some astronomical closing ratio. They're great when you get one, but honestly, your company didn't hire you for those. That's just a freebee.

Alright, now you have a few leads. You will need to keep track of them for a number of reasons, and like with most things in business, the more information you have, the better chance you have of improving your results. So, first off, where did these leads come from? It is vitally important to both your advertising and your closing ratio to know where these leads originated. If after going on ten leads from your direct mail campaign, you find that you only closed one, but out of the ten that you got from the radio spot you closed eight, what does that tell you? Where are you going to spend your money on advertising in the future?

Then you will want to look at how qualified the leads are that you are getting. Did the direct mail ad say something like "we don't care if you have a job or not, buy today!"? Maybe the radio ad was being run on a financial station where the average listener has a household income of over $200,000. These are important factors to decipher. You will also want to be careful about what is being said when the appointments are being set. Are all decision makers going to be present? Are they expecting to make a purchase in the next thirty days? A warm lead that isn't qualified isn't going to help you put food on the table.

If you come across a lead that is qualified, but tells you that they aren't expecting to make a purchase for another six months or so, what will you do with that lead? I've seen countless salesmen throw the lead away and say "they'll call back when they're ready". I don't understand, don't these guys expect to still be selling in six months? Maybe they don't, but for those that do, they should be adding that lead, with a long list of notes, to a tickler file of some sort.

A tickler file is a file kept by salesmen in chronological order, that is to say that it is a file organized by date. When you have a lead that you've spoken to or met with, but won't be ready to buy until September, you take all of the info you have on them and put it into the August section of your file. Then when August comes around, and business is slow, you have someone to call that you know is going to buy soon, and it's already a warm lead. What's better than that?

I quickly said that you should take all of the information that you have on them, and put it in a file. Well, what exactly does that mean? Of course, their name, phone number and address, but what else? Yes, there's more, and you're not going to remember half of what you think you will. No, you won't. The information that you'll want to record is more than the details are of what they were looking to purchase. Imagine if, when you called Mrs. Soccermom in six months, she doesn't remember speaking with you. Now what? Your warm lead just turned ice cold; but what if you were able to say something like "so how did Katie do in the finals in June?" Mrs. Soccermom will be amazed and flattered that you took that much interest in her family, and even if she doesn't remember speaking with you, she's not going to admit it now.

There are all kinds of things that you should be picking up on and noticing when you are in someone's home or business, or even just speaking to them on the phone. Use a little discretion here, though, you don't want to come across as creepy. But, what type of sports they like, or their favorite team, maybe that your kids go to school together, or that they told you that they were going on a trip to the Grand Canyon this summer can all be things to remind them that you hit it off last time, and that you were someone that they wanted to do business with. And while we're at it, when you do call them to set up that next appointment in August, remind them that they *told* you to call back at the end of the summer, and that they would probably be ready then.

What we're talking about here really, is taking a lead and turning it into a prospect; a prospect being the next step in the sales process. A prospect is just a lead that you have made contact with and have a little more information about. Hopefully for you, these are beginning to add up now, and you've stopped throwing your future customers in the trash, but in doing so you will need to become organized. Salesmen are notoriously disorganized, but there is help. Certainly, you could simply use an expandable 12 pocket folder for your tickler file, and some other method for the rest of your leads, but I'm going to suggest that you stop using the old donut box in your car, and pretend that you're more professional than that.

In today's wonky world of electronic paperwork and other mystical wonders, there exist a good number of sales assistants ready to make a home on your laptop. One in particular comes to mind, and that is ACT. ACT is a program designed to do just what we're talking about, namely to keep track of leads, prospects, and really all stages of sales. Surely many programs are available, and I encourage you to find one that you're comfortable with. As a last resort, at least create a spreadsheet for the pertinent info, but really you can do better.

Just quickly, the other side of this proverbial coin is the idea that old leads never die, they just go in the pile of "maybe someday". I used to do this, back in the day. I would never completely give up on a lead, and somehow those old leads were like a security blanket. I thought that if I ever was *really* desperate, I could revive them and sell them something. Eventually I realized that I was just fooling myself. So while you don't want to throw out a lead that isn't ready to buy, make sure that whatever you do save has actual potential to turn into a sale at some time in the future. Be honest with yourself. Sometimes you'll come across a lead that is going to buy, but it's never going to be from you. Who knows why, maybe you remind him of the bully that used to take his milk money in second grade. Don't take it personally, it just wasn't meant to be. Give the lead to someone else to try, and move on. There's an awful lot of equipment out there waiting for you; don't get caught up on that dead lead. "I'm done wit *dis* guy" (you remember the movie, "My Cousin Vinnie"?).

Take-Aways

- Hot leads don't require your skills, that's cheating
- My barking dog is distracting me
- Qualify your leads
- Keep good notes on your prospects
- Make a tickler file
- Find a sales lead program that works for you

Your customer is who?

So, tell me, who is your typical customer? Do you know? Somebody in the back said "he's the guy that's going to give you a bunch of money". Yes, that may be true, but let's take a closer look at who these characters are and what they're made of. I want to know what your average customer wants, and why they ended up buying from you. Your average client didn't choose to buy from you by accident. And, while it's true that *some* of them decided that price was the most important factor, in reality, and through research, we see that most people said that price was less important than some other aspect of the deal.

I know that a bunch of you don't believe what I'm telling you, but the facts are the facts. If indeed price was always the most significant reason for them to say "yes", then brands like Cadillac and Prada wouldn't exist. Just because *you* do something a certain way (your buying decisions, in this case), doesn't mean everyone else, or even most people, make the same choices. Luckily for the rest of us, they don't. It takes all types of people to make up this crazy world of ours, after all; how will you make it work for you?

Take a few minutes and go back through the last twenty or thirty jobs that you have sold. Do it now; turn on your computer or open up the shoe box and take a look at the items that your customers have purchased from you. Not the things that you gave them for free, but the things that they actually paid for. This should give you a clue to who you are attracting. Are you installing a lot of high end, brand name equipment, with all the bells and whistles, or is it all no name, cheap-as-possible type stuff? We're going to get into this more later in the book, but in most cases, this says more about you, than your customer.

Earlier, we discussed advertising a bit. Through the advertising that you do, you can control, to some degree, what types of customers you attract. The content of your ads, that is, what you focus on or talk about, will determine what type of person picks up the phone to call you. If your ad says "Get a new furnace today from the <u>Budget Bastards</u> for $999.00!" you're certainly not going to attract many homeowners that want to show off their new purchase to their friends on game night, or take photos in front it for the family Christmas card. They don't see this purchase as anything more than a necessary waste of money. Quality of life, pride of ownership, and peace of mind don't have a place in their vocabulary. These customers are "price-only" people, and they make up a good part of your audience.

The other end of the spectrum should be pretty obvious to most of you. Many customers are not so concerned with the price, believe it or not, as long as you can solve their problem(s) and make them feel as though they made a good decision. They know their house is too dry in the winter, and the room in the corner is always too hot or too cold. They've done their research, and have their budget in place. Others, very specifically just want the best of the best, because that's how they do everything. They want the one with the best Consumer's Report's rating, lowest decibels, and highest SEER, whether it makes sense or not. This guy has the perfect lawn, shiniest red car, and wife with the boob job. It's all about the show. Think about it; you know one or two of them yourself.

So, what did you find out? Which one of these customers do you sell to most often? Is that who you want to sell to? I'm not saying that you have to focus on one or the other, that's up to you. But the first one, Mr. Price-only is much more difficult to make a living with, and probably not as satisfying to have as a customer. You can, without a doubt, influence which of these examples want to do business with you by your advertising and your sales pitch. Do you really want to just attract the guy who just wants the $999 furnace, and even then is going to haggle for every penny of it? You have to decide which customer you're most comfortable with, and what you want to sell.

Regardless of what you decide and the advertising that you run, you're still going to get calls from the guy who says "Don't waste your time with the sales pitch, I know what I need. Just give me the number and if you're the cheapest, you'll get the job." Maybe that is the type you're looking for. You don't need to do a big sales pitch, and if you do get the job, you're in and out quick. Personally, you can have him, but maybe that's what you decided you do best. There're lots of those customers out there, and if that's what you want, you should never run out of work. Unfortunately, if that's the only kind of business that you want to focus on, I don't think you're going to get much out of this book, or be in business long, but who knows.

I want to be clear here, in that, you can make a good living and do very well with which ever type of customer that you choose to work with, regardless of any implied bias on my part. If you do your research, really investigate your results, and are honest and objective about what works and what doesn't, you will do well with any group. A business with a focus on one segment may look completely different than a company that plays to another segment, but that's ok; they all need us one way or another. I've noticed that we don't all root for the Steelers, either. You'll learn.

So, then, it's important for all of us to know who our typical customers are for a number of reasons. For one, without knowing whose attention that you're trying to attract, you won't be able to put an effective advertising program together. Even if this is all new to you, or you've never given this stuff a shot glass worth of thought, you must realize that the "price" guy, and the "20 SEER, IAQ" guy are going to respond to much different types of advertising.

Another reason to know who your "normal" customer is might be the resources that you carry in your salesman's toolbox. Think about the type of financing that you offer, if any. Chances are good that if you do mostly residential work, and then you land a small commercial job that you feel comfortable doing, you'll worry about getting paid because your finance company won't underwrite that type of work. If you know ahead of time that you may work in that arena, then you can prepare for how to help your customer pay you; more about that later.

I know, for instance, that you are involved in the HVAC world (duh), and have decided that you want to grow your business, make more money, and become more efficient in your sales process. How do I know that? Because that's how I market this book. Just as you should have more interest from your customers for indoor air quality products if you include that in your marketing. It all ties together. The same goes for anything that you put in your advertising. If you say you have exceptional service, you'd better, because that's what the people that respond to those ads will expect.

As we discussed earlier, you need to know what the goal is, or more specifically, who your target customer is. Where you end up on this topic has only to do with you, and there is no "right" way. What I mean is, what are you comfortable with, and what do you believe in? If you are the type of company that believes that nobody cares about the name on the box or anything else besides the price, AND, you have been successful at being the "budget" contractor, then more power to you. That's what you should do, and that is the type of clientele that you should pursue.

Obviously, if you prefer all of the latest high tech gadgets, and the most efficient, highest rated, shiny, best-of-the-best, then you know that you're going to spend most of your time on the other side of town. You should also expect that you will probably need to spend more money on advertising, have nicer trucks and equipment, and focus a lot more on your dog and pony show compared to the previous contractor. As such, don't expect to get away with the same "number scribbled on a napkin" estimate either. You've got to come across as a professional, which takes more time and effort. To be sure, your jobs will also take longer to complete, and the customers will expect a lot more, but the number on the check at the end will be much heftier, too.

These are two of the possible customers that we come across every day, but who else could you make a living selling to? Well, actually, there are quite a few other customer types to play to. You could spend most of your time in new construction. You could do all commercial or industrial work. Maybe you specialize in hydronics work, or radiant floor heat. You could make a good living selling service contracts to fast food restaurants changing filters. I know companies that only do rooftops or ductless splits. There are companies that specialize in historic buildings and others that just do pool areas. Then there are green houses and farm buildings, or medical facilities. The list is almost endless.

This chapter isn't meant to try to change who you do business with. I want you to discover who you currently work with, and who you feel most comfortable working with. What do you believe in? It doesn't matter if it's "A" or "B"; you just have to know which one it is so that you can best prepare yourself for success.

Take Aways:
- You need to know who your customer is
- You can make a living with any of them
- Convert to the world of Black and Yellow

What are they thinking?

Regardless of what type of work you are involved in, people are people, and we all have our misgivings, biases, and idiosyncrasies, right or wrong. As humans (this may be a stretch for some), we all seem to instinctively think that standards, basics, and common sense are common. I mean, for instance, everyone would rather keep their shoes on when walking into a stranger's house than take them off, right? You don't know how clean they are, or where they put their used hypodermic needles. But some are offended if you *don't* take them off. How dare you bring dirt from the street into their home!

To complicate things more, we all seem to communicate differently. Just because I said "blah" doesn't mean that's what you heard. Nowhere in life is this more obvious than in a relationship between a man and a woman. The longer that I am married, the more that I realize that my words, sentences, and intentions, are often received waaaay differently than I meant. I'm not accusing or blaming anyone (well, actually, her name is Candace), that's just the way we are.

Throughout our lives we learn to interpret words, tones, expressions, faces, body language, and many other kinds of signals. If your friend tells you that the chili that you made for dinner was great, it makes you happy and you feel good. But what if he said the exact same words, but just made a little smirk, and looked at one of the other guys when he said it. It *could* take on an entirely different meaning; perhaps a sarcastic one. In our attempt to communicate with the world around us we use the knowledge and signal interpretation acquired from our entire lives. The only problem with that is the fact that whoever you are speaking to has acquired slightly different knowledge and signal interpretation than you have. You are speaking a slightly different language, and neither of you realize it.

What do you think; any chance that could cause a problem in sales? Noooo, it'll be fine, you say. Uh huh; like when you pull up to an estimate for the first time and the customer is standing outside looking at their watch, tapping their foot because you're a half an hour late for the appointment. "Hmmm, that's not good" you think, "but I told them that I would be here around eight". Except that "around eight" to them means 8:00; not 7:59; not 8:01. I'm not kidding; you know who I mean, you've met that guy. But to you, "around eight" means "I'll leave the shop about 7:45, grab a coffee for the ride, and be there in a few; it's only twenty miles from here." Sounds like a quarter to nine to me.

There is a *lot* of psychology that comes into play in the sales world, as you will continue to see throughout this book. Much of it relies on your ability to properly communicate your thoughts and intensions, and likewise, your ability to figure out what your prospects and customers are really trying to tell you. Everyone thinks that they know, and are already pretty good at this, but I'm going to tell you that you don't know what you don't know. You aren't nearly as good at this as you think. I realize this more as I get older, and how difficult it is to have someone understand the point that I intended.

Timeout: I travel a lot for work, and I just arrived at my hotel. I walked up to the counter as usual, and waited while the clerk finished up with the guest ahead of me. There is a small vestibule at the front door with an automatic sliding door on each side. So, as I'm standing there waiting, the outside door opens, but no one is there. Then in strolls a pure white, little mouse; not in a rush, just walking around, checking the place out. The door closes behind him, and he continues as if he is the new manager. I turned to the guy standing next to me and said "did you see that?", "I sure did" he said as we both chuckled and watched. Suddenly the door opened again, and the mouse left just as he had arrived, not to be seen again. Yup, things that make you go "hmm."

Let's continue. We now know that we don't speak the same language as anyone else on the planet, and no one understands each other, just like you thought when you were thirteen. What else could go wrong? I'm glad you asked! (And by the way, this is all positive stuff, here. I don't want to hear any negative Nellies. If I tell you that you suck at this, and you don't know anything, it's all good. Rainbows and unicorns, you hear? Don't worry, you'll get it; remember the guy in the beginning with the stupid hair?)

Next, we're going to talk about what the customer is thinking and feeling when they see you, and when you come into their home. I want you to put yourself into their shoes. No, not literally, you big pineapple! I mean, imagine if *you* came to your house for an estimate; what would you think of you? Are you a mess? Are you scary looking? Are they trying to figure out whether you're the HVAC salesman, or the escaped psychopath they just saw on the news? Selling anything is difficult enough without complicating it because you look like a drunken Pig Pen that just finished plucking the Thanksgiving turkey, wearing your Dora the Explorer slippers. Be aware of yourself, and please… buy a mirror.

It should go without saying that you need to be *very* aware of the customer's feelings, and the fact that you are in their personal space. Sure, you do this every day, that is, every day you go to a few new homes and meet new people. But, your customers don't. You're a stranger in their home, and that is very uncomfortable for them, even scary for some people. This is their sanctuary, their safe place. You know, it takes a certain type of person to be in sales, and because of that, you may not be able to understand what they're feeling. That's OK, but you must respect the fact that they may think of you as an invader. They may actually consider you threatening; seriously.

I can't stress this enough, and I'll probably repeat it a few more times. Most salesmen, especially the newer ones, do not realize how uncomfortable it can be to have a stranger come into their home, wander around, and look in every single room. People will be embarrassed, feel judged, feel vulnerable, and a bunch of other emotions. You must take this to heart, and use this understanding to help make the customer more comfortable with you.

Right from the beginning, ask permission for everything. I mean, really go out of your way to be non-threatening. Ask if you can put your jacket here, or is it ok if you put your bag over there. It may sound silly, but you want them to feel as if they're in control, and that you pose no threat. This is especially true when measuring a house, or trying to figure out where something may go, like a diffuser, and you need to walk around and go into different rooms. Always ask permission before entering another room and be sure to knock if the door is closed. This will show them that you are respectful, and will help build their trust.

I'll never forget the time that I went to do this air conditioning estimate and the husband and wife met me at the door and invited me in. We sat down at the kitchen table and I went through my regular presentation. They chose a system, agreed on a price, and wanted to get stated right away. I explained that I would need to do a load calculation, and that in doing so, I would need to see and measure each of the rooms in the house.

So, off went the three of us, room by room, as we talked, and they asked questions. "Ok, the first floor is done, can we continue with the upstairs?" I asked. "Sure", they answered, and led me to the second floor. After we measured the first room, I didn't notice that the wife disappeared. I went to the next room, opened the door, and then heard a loud shriek! It was the wife, who had decided that this was a good time to change her shirt. Wait, what? Didn't she know we were going there next? You would think, but she didn't seem to get the memo. Maybe, that was just a thing that she was… nope, she was genuinely angry that I had opened the door. (Sigh) Just knock.

No beat missed here. "But that stuff only applies to residential customers, right? I only work in the commercial markets, so I don't need to worry about that." Oh, contraire, my friend. Peeps are peeps, and even if you're not working in their most sensitive of places, that being their actual homes, you will still be subject to their prejudices and reservations. To be sure, the work place is less threatening, yet they will remain on guard until you prove them wrong.

Ugh. OK, so one thing that we know for sure is that, just like us, your customers have preconceived notions about much of this sales process. The same way that you "know" that this guy isn't going to buy the UV light, he's too cheap. Your customers think they "know" you and your workers, and how you/they will act in their home. They "know" that you're going to make a mess and rip them off, and they "know" that your workers are going to break something, and that list is as long as my leg. Face it; your customers are afraid of you. Yes, little 'ol you, with your big bushy beard and Sponge Bob tattoo. Let's talk about what they're afraid of.

I'm not even sure where to start on this subject. The list is long and may be completely unjustified, but you still need to deal with it. All throughout our "normal" lives, people will make judgments about all of us, and decide whether or not they like us, or would include us in their circle just by the way we look, dress or act. Like it or not, that is just human nature, and many of these decisions are made subconsciously, never giving you a chance to refute them. You do it yourself every day. You see someone and think you know how they "are", like the girl you thought was cute at the bar, but wouldn't talk to because she "had an attitude" That's all you, Jack, although, in this case, she *wasn't* going to talk to you. Sometimes people don't give you a chance to make a first impression; that's just life.

Beyond the uncontrollable, you're still left with a laundry list of things that your prospect is worried about. On that list are things such as cost. Naturally, they're wondering how much this will cost, and how will they pay for it. For many people, this can be overwhelming and even paralyzing. Then you have things like, are you ripping them off, are you telling them the truth, is the system efficient enough, is that a good brand, is it the right size? Do we really need that filter, should we get the humidifier, will that really help Jimmy's allergies? Will they do the right job, and how will I know if they don't? What if they break something; who will fix it, do they have insurance? Is that guy a crack addict? Is he going to steal from me? Is he a sex maniac?! What are the neighbors going to think? AM I MAKING THE RIGHT DECISION?!!!

Sounds to me like you may be dealing with a stressful situation, no? Now, to be clear, not every customer is ready to jump off the bridge, but some are, and you probably aren't going to know which are, and which aren't, so you have to treat them all like jumpers. Don't take this lightly; these people are going to buy from who they feel most comfortable with, not necessarily who's got the cheapest price; that's *your* preconceived notion.

Then there are the people that probably didn't want to buy anything now, but the #@*&&^$#% thing broke, and they have no choice. This is not a happy occasion, like getting central air conditioning for the first time. They're under a lot of pressure, and they're scared! Will you put them at ease and be their hero, or make their anxiety worse and be the hated villain? Your choice.

Yup, there is plenty to know about your customer, but they probably won't come right out and tell you what they're worried or concerned about. Casually address the common issues as you go through your presentation, like insurance and drug testing your employees. Above all else, you need to build their trust in you, and make them feel that they are making the best possible decision.

Take-Aways:
- You're scary (to them)
- You have preconceived notions, and so do they; you're both wrong
- Ask permission for everything
- Make them feel that they're in control
- Assure them that they are making a good decision
- Always knock first

When in Rome,... Don't Speak French

Do I need to explain this one? Literally, you should speak their language. Of course, I am speaking about your customer. You would think that this could go without saying, but no, it can't. I once knew a salesman who was originally from someplace like Albania. He would go on an air conditioning estimate, and throughout his time there, would keep bouncing back and forth between English and Albanian. He would speak to the people in English, then, suddenly he would use expressions from his native language, and didn't think anything of it! And, no, he never explained any of it, as if he expected them to understand it all. I suppose his game plan in the house was to make his customers feel as *un*comfortable and alienated as he could. I think, back in the old country, he would just beat them with a stick.

Ok, just to be sure that we're all on the same page, we're all going to stick to one language at a time, right? Good, let's move on. The first part of a sales meeting begins with attempting to remove, or at least reduce, any barrier that may hinder the production of a sale. This includes anything that keeps the customer from trusting you, creates distance between you, or basically says "He's not like me". In other words (speak English), you want to show your customer that you have as much in common as possible. Yes, I suppose you could say something like "Hey, I have a wart on my nose just like yours!", but that's not exactly what I had in mind. You could, however, say something like "Your tomato plants look like they're doing well, what's your secret?" Here you have complimented his success on something, and implied that you are a fellow gardener without actually saying that you are. You may hate any food that isn't filled with 90% preservatives and chemicals and comes out of a factory, but you just made a connection with your customer and by doing so you've lowered a barrier and increased your chance of closing the deal.

Already I can see that some of you are not convinced of the importance of this step. Some of you are "all business", and don't see the need for this non-sense. Well, I will at least say that how *much* of this you will need or want to do will depend on many factors, such as your personality as well as your prospect's, how well you hit it off in the beginning, and the degree of necessity of the item being proposed. But regardless of each of those factors, the likelihood of closing the deal now or in the future will increase dramatically for all salesmen if a personal connection is made between the two parties.

To backtrack just a little, the first time you speak to a potential customer, including on the phone, you should immediately search out common ground. Think of yourself as Sherlock Holmes, intently digesting the surroundings for clues of what defines this person's life. A Pittsburgh Steelers bumper sticker on the car in the driveway, a picture of a little girl in a soccer uniform, even a pair of Merrill running shoes, anything that you can start a conversation about, and say "me, too!" If on the first contact with a prospect on the phone they ask you to call back later because they are on the way to a school play, make a note of what you have just learned. Your client likely has children, they are involved parents, and their kid(s) may have been in the play; should be plenty of conversation starters there. The trick is to simply remove the walls between you two, and make a connection.

You may have heard somewhere along the line in a sales class that, if you want to get someone to like you, get them to talk about themselves, which plays right into what we were just talking about. Naturally, this applies outside of work, as well. If you don't already incorporate this into your everyday life, you may be surprised at the results. In coaching my daughters through their shyness, I would explain that this technique solves two problems. First, all you have to do is ask the right question to get them talking, then you can be shy and just listen as they ramble on, and secondly, you'll have made a friend, because you expressed an interest in them. It really is as simple as that; you just need to find the right topic.

The key topic is likely right in front of your nose, if you're at their house, that is; just find where their passion lies. Look at the pictures on the walls, the furniture in the house, the car in the garage, the books on the table; the possibilities are all over the place! Be careful, though; remember that you're trying to build their trust in you. If they have an overwhelming affiliation to a certain political party, for instance, don't place yourself in the opposing camp thinking that this will be a fun and friendly debate. That may expose too much passion, and drive a permanent wedge between you, killing any chance to ever close a deal.

So then, the first order of business with a new client is to find common ground, and get them to like and trust you. All people want to buy from people they like. There are countless numbers of sales lost of every kind, due only to the fact that the buyer didn't like the seller. Despite the price, brand, workmanship, or anything else, a solid connection can make or break a deal. This goes double for the trust factor. As you have more in common with your prospect, you'll see that they trust you more for some reason, too.

Now, you're in the house, everyone has been introduced, and it's been established that you each have kids in the varsity tough-man rugby program in the local grade school. You discovered that you each stayed in the same hotel in Hawaii when they had the mumps epidemic, and you both like to grow Mongolian bananas. Great! They like you, they trust you, and they invited you to Grandma Jeanie's birthday party. Now what? What do you mean "now what?"? Now you do your thing! Start fact-finding, and do all of that presentation stuff that you do! Wait, I'm jumping ahead again, let's review your presentation now.

I'm sure that you already have some sort of initial presentation that you do when you first sit down with a prospect, whether you realize it or not. This is like doing the company's version of an introduction like you just did for yourself. Before these people buy any equipment, they need to "buy" you, and then they need to "buy" your company. Think about it; there are places you won't shop because you don't like the owner or manager, right? It's the same thing.

Just like you revealed yourself to your new friends, you must unveil your company. People want to do business with companies that they relate to, that they trust and understand, and want to know that they're not a "fly-by-night" operation and will be there if something goes wrong. So, your next piece of the sales puzzle is to put their minds at ease by telling and *showing* them who" ABC-XYZ HVAC" really is. The best way to do this is with a professionally bound book, or perhaps a video detailing all aspects of your operation. The more you can document here, the easier the close will be later. Show them pictures of all the employees, and the buildings and trucks. Give them a history of the family business, and the community projects they're involved with. Try to find the same type of common ground for your company as you did for yourself.

Having a few letters from satisfied customers in your book for them to browse through is also very powerful. To be completely happy with someone's work is one thing, but how often have you taken the time to actually write a letter to that effect? I thought so. How about going one step further than that? What if you actually told them about the time Louie was up in the attic, lost his balance, and put his foot through the ceiling. No, I'm not trying to make you look like a fool, just wait, there's more. Of course you can't stop there, the point of the story is to tell them how you had contractors over there right away to fix the ceiling, how you painted the entire room, and then sent over a cleaning crew to clean their entire house. If that's not what you do, then leave Louie out of it, but if you do, your customer will be sold on your company, and then it's just about how *much* are they going to buy from you, not "*if*".

"Great! Let's go close the deal, do the work, and get paid!" you say. Alright, yes, we're getting closer, but we need to discuss all of the fact finding and estimate stuff too, but even before that, let's talk about "jargon". (Man, I'm never going to get done with this book.)

When in Rome,... Don't Speak French

Jargon is not a type of fish. Jargon is another method used by some salesmen to drive their customers away from buying. It is a thoughtless procedure by which one uses words of the industry or of a technical nature that only those engaged in the field professionally would recognize. You know how when you go to the doctor, and describe your problem to him he starts using all of those words with ten syllables but only two vowels in them, before he says that it's going to burn when you pee for a while? Yeah, well, I'm talking about those big words in the middle, just like if you were to start talking about CFM and face velocity of a diffuser to your customer, they'd be lost, too. Listen, you spent a bunch of time making a connection with your customer when you first met; don't waste it all and push them away but talking about things that they don't understand.

It's important that you find just the right balance of technical and non-technical when speaking to your customers. Too technical and you're talking over their head making them feel confused and unable to make an intelligent decision. Too fundamental and they'll be insulted that you think they're too stupid to understand. Either way, you'll squander all of the ground work you laid in the beginning, and again risk the sale.

To further complicate the issue, wait, let me change that. To make your sales call even more invigorating and exciting (staying positive), you'll never be quite sure who you're speaking to. I tell this story often, because it made such a huge impression on me, early in my career. If you've heard it before, just be quiet and don't give away the punch line.

I was given a lead for a gas furnace replacement in a nearby town. I prepared my literature, as always, gathering brochures for my good, better, best scenario, along with a few options for good measure. I pulled up to the house right on time, rang the doorbell, and waited. A minute later, after hearing some commotion in the background, a girl or small woman answered the door; I couldn't tell how old she was, either twenty eight or fourteen, I guessed. I introduced myself, and she invited me in. The place was a vicious mess, with toys and diapers everywhere. Clearly, she was the mother, I got it now.

She pointed to the furnace, and off I went. A few minutes later I emerged from the basement, and I asked if we could sit down to discuss my findings. She was polite, although obviously exhausted, and showed me to the dining room. Apparently, her husband wasn't going to be home for a while, but she said that she could handle this. A "one legger" as I would call it when only one decision maker was present was usually the kiss of death, ensuring that no deal would be closed that night.

However, I pushed on, figuring it was good practice. I went on to tell her about what I had found, and what the differences were in each of the three furnaces I suggested. I explained what a two stage gas valve was, and how that would help save fuel. Then I painted a picture of how the variable speed blower would run slower but longer to create a more even temperature around the house and help filter the air. This went on for about twenty minutes, then, at the end, she thanked me. She said that she appreciated me not speaking down to her because she was actually (here it comes…) an HVAC engineer, and was very familiar with what I proposed. She bought it all, and I never met her husband.

The moral of the story is that it is extremely important to strike the right balance of "technical speak". Learn to read your customer's body language. If while you are describing something, they start looking away, or seem to be uninterested, stop and ask a question, or do something to get them engaged again. Ask them if they understand what you said, or if they would like you to go over it again. If you built a connection before, this shouldn't be a problem. Learn to walk the technical, non-technical line.

So, to wrap it all up in a nice and simple package, be like them, and you'll greatly increase your chances of closing the deal. Make them believe that you have stuff in common, and don't blow them away with technical stuff unless they want it. You can ask them if they want to hear all of the technical stuff, and if they do, lay it on them. But for the most part, they won't care. They just want someone that they like and can trust to fix their problem and make it all go away.

Take-aways
- Search out common ground, then use it!
- Get them to talk about themselves
- Get them to "buy" you first, then the company, before anything else
- Don't assume anything
- No jargon allowed
- Don't get too technical unless they want it.
- The goal is to get them to trust you to fix their problem, and feel good about their decision

Fact Finding = Listening

If you are in the habit of going to an appointment and trying to sell something without first finding out all of the pertinent information from your customer, you should just go stand in front of your boss, and fire yourself now. By pertinent information I mean, everything from their problems to their budget. You're not doing right by your clients, and you're giving the industry a bad name. People will see right away that you're just there to make a quick buck, regardless of what they really want and need. You don't need me for that. But if that's what you were doing, and you realize that you need some rewiring, keep reading and we'll keep that our secret.

You simply cannot do a proper job without asking lots of questions, and that takes time. Don't expect to increase your sales and/or your closing ratio without first increasing your time in the house, by the way. Good leads are not easy or cheap to come by, so don't waste a single one. If you spend the time to do the math, you'll probably find that you (or your company) spend literally hundreds of dollars for each new sales or service lead you get, so pay attention.

I see service managers all the time try to rush their technicians through a call because they are backed up, and have ten more on the board. Their priority becomes getting the calls off the board rather than doing a proper job for the customer, and being profitable for the company. (Yes, they're both important). I can't tell you how many times I've asked service managers if they made any money at the end of a really busy day, only to find out (to their amazement) that they didn't. Service departments are profitable when they stay in the same place for extended periods of time and repair or install things, not by driving around collecting service call fees or trip charges.

Sales leads are no different than service calls in this regard. The goal (another one!) isn't how many leads or service calls you can get to in a day, but how much money can you make while doing the right job and creating a satisfied customer. If you can only get to one a day, but make a good amount of money on every one, you will be in high demand and highly compensated, my friend. And, I'll take a stand here and say that a good salesman's sweet-spot is *two* leads a day, *maybe* three once in a while. But with more than two, you usually can't prepare or follow up properly, if you're doing a good job.

So, what exactly is a good job, you ask? Well, to me it is a very satisfied customer that would happily use your company again and recommend you to their friends. This is a customer who feels that you solved all of the problems they told you about, and probably some that they didn't realize that you could. They are excited about how much they will be saving on their utility bills, and very comfortable with how they agreed to pay for it. The work was done quickly and with few, if any, mistakes due to your great communication with the installation department. You asked for a few referrals, and got three, and they look forward to seeing you again in the spring when they put that new addition on the house. And, of course, you got paid as agreed, and the gross profit on the job was exactly what you had planned. You both win!

Once again though, you cannot do a good job without knowing what their problems are, or what your clients want, or at least, think they want. I say it that way because your potential customers really don't know what they want. After all, they're not in the business, and probably don't know 95% of what even exists in our world. Hell, I can't even keep up with all the new technology and products that come out every day; surely they don't have a clue, despite the fact that they did do their research on the internet. They are relying on your expertise to guide them through the maze of HVAC craziness to make a good decision, even if they don't say that to your face.

Thinking that you know what's best for them is arrogant and obnoxious, so let's not do that. Remember that you want them to feel that they are in control of this process in order for them to relax and not feel threatened, and so that they will listen to your suggestions and advice. So, the first thing to do now is to explain the process that you will take them through so that you have enough information to recommend appropriate solutions. Explain to them that even though you've been doing this for XXX years, without going through this process, anyone would just be relying on guesswork, and that their comfort, health, and money are well worth the time it takes to do it right. You refuse to cut corners, unlike so many others out there.

Now, think about what you just did here. If anyone else comes to give them an estimate (and make no mistake, others are coming), and doesn't take the time to go through the process that you are about to embark on, you have established that they are cutting corners, and don't care much about your customer's comfort, health and money. You have effectively beaten down most of the completion without mentioning a single name. Don't worry, they will remember and notice that you are/were much more thorough than the other guys.

OK, what's next? You could just start asking whatever questions you think of that seem relevant, since you've done this plenty of times before, right? Go one step further and bring a pad and a pen and write down the answers and you'll look that much better. But, at this point you have already set the stage for what promises to be a fairly comprehensive inquisition into the homeowner's wants, needs, and unresolved issues. By just shooting from the hip, as it were, you are putting forth an anti-climactic approach that frankly lets you slip back into the unprofessional category of your common peers. You must continue down your path of certified home comfort expert, and play the part by producing a process that is calculated and second nature to you.

What!?! FILL OUT THE FORM!!! Yes, fill out the form. You want to come across as having a regular, standard, calculated process to analyze. I'm not saying you should *"act as if"*, at all. I'm saying that you already do this every single day; you should have a standard, regular form that you are familiar with, because you do fill them out every single day. It doesn't matter where you get this form, make it up yourself. It will make you look professional, and just as important, it will keep you on track during your interview so that you don't forget anything (like you usually do; tell the truth.)

This form, or questionnaire, should have all of their basic info like name, address, etc, all the way into subjects like "which rooms are too hot/cold, who has allergies, are there any noises that really annoy you?", and so on, but it goes way beyond that, too. Most of this stuff you already know, but just never put it all together. You need to measure things, and do a load calculation. You need to check wire sizes and note what brand of electrical panel they have and whether or not they have enough space in it for your project. Are they interested in financing, what brands have they heard of, what other projects would they like to get to in the future, are a few of the things that should be specified on your form. This will be a long list once you really start thinking about it, and it should be. To do the right job, you need lots of information.

I don't think we need to discuss each individual question that you decide to ask, but here is a list of some that I think are important.

- What do you hope to accomplish with this project?
- What ideas do you have?
- Do you know anyone that has had this done?
- What were their experiences? (Were they happy?)
- Did you do any research?
- What did you learn?
- Did you have a budget in mind?
- How do you plan to pay?
- Are you aware of our financing programs?
- How long have you lived there and plan to stay?
- Are you aware of any permits that are needed for this work?

- Rate the following topics from 1 – 5 : efficiency, aesthetics, status, brand, carbon footprint, space used, quietness, color, etc.
- What future projects are planned?
- Are you concerned with rebates?
- Are you interested in reducing or eliminating dust and/or odors in the home?
- How does your existing system perform on the hottest and coldest days of the year?
- Would you be interested in a backup heating system?
- Do you understand about mold growth on an air conditioning coil?
- Do you have carbon monoxide detectors in your home?
- Do you know that they need to be replaced every five years?
- Are you interested in doing any of the work yourselves?
- Do you run a business out of the house, or do a lot of entertaining?
- How many people at one time?
- Do you do a lot of cooking, or do any catering?
- Is there an older person, or someone that has special needs? (High temperature room, etc)
- Tell me about any special humidification needs (antiques, artwork, etc.)

This is just the beginning, and don't apologize for it being long to your customers, it's for their benefit. Different markets will, naturally, have different questions and concerns, and you will probably want to leave an area for a sketch of the house, if needed. This is as much for your benefit as it is theirs, so design it however it works best for you.

I mentioned a load calculation quickly a paragraph back, or so. Let's talk about that for a minute. Are you doing load calculations on all of your new and replacement systems? You really need to. Most, and I mean almost all, heating systems are way oversized, and may be located in a house that is quite different now that when it was installed fifty years ago. More insulation, better windows, and other types on new technology contribute to more efficient homes, requiring you to do a new Manual J for the proper diagnosis, not to mention that more and more municipalities are mandating them now, even for a swap-out.

Beyond that though, most of your competition *isn't* doing them, and this will again put you out front in the race to do the right job for the homeowner. Let them know how involved it is, and don't down play how easy the new programs are. Explain to them how it calculates the direction of the sun, and the R values, the construction and the occupants of the house to come up with the needed BTUs and CFM for each area of the house, blah, blah, blah. Don't be overly technical, and you will impress them with the results and your commitment to their comfort.

And by the way, certainly consider upgrading your presentation and fact finding to a paperless iPad or laptop. Once again, this just puts you way ahead of the guy that comes in the house, grunts as he hobbles down to the furnace room, and comes back in five minutes with an estimate scribbled on a Snickers wrapper, and says "call me if ya wanna do it". You could email your questionnaire and estimate to them before you leave their dining room table, along with all the brochures and references they need to reach a conclusion.

Most of us will have to go up in an attic, or go someplace during the visit that the homeowners don't want to go to. In other words, they'll just say "it's over there", and expect us to go and come back as we need to. At that time, I like to have a form ready that they can fill out themselves, like a "rate these from 1 – 10" thing, or something that is fun, yet makes them think about something that they didn't know they wanted like a UV light, or a hot and cold faucet in the driveway for washing their car.

Another thing that I want to remind you of is the fact that often times when your customer tells you what's wrong, they're wrong. It's not their fault, they're not telling you the wrong thing on purpose, they just don't know what they don't know. Take whatever "facts" that they give you, and do your own investigation, with your own conclusions. Remember that you want to solve their problems, not fix what they think is broken. They want the water to stop leaking, they don't understand, or care, why that has anything to do with the air filter.

And lastly, unless your customer has specifically told you what they want to spend, get rid of every single pre-conceived notion about their finances. Ask questions about what they want and need, all of the options that they might want, the highest efficiency rating they would be interested in, the twenty year worry free warranty, all of it. Don't worry about the cost now, just include everything that they asked for. We'll discuss the proposals, paying for it and closing the sale soon.

<u>**Take-aways**</u>
- You're the man!
- No more than two leads per day, average
- Ask lots of questions, then listen
- Use a pre-printed form, don't wing it!
- You can't satisfy them if you don't know what they want
- Be much more thorough than the other guy
- Do a load calculation
- Don't believe what they tell you, find out for yourself (don't tell them that, ya big dope)
- Don't worry about the money here; assume that they can afford it all

A Decent Proposal

The tension is starting to build, I can feel your heart beginning to race from here; will he be able to close the deal? Will they go for all of it? Will they sign NOW!?!!! Hold on there, Johnny Rocket; will they sign *what* now? Put yer pants back on cowboy, we're still not there yet.

Now then, you've established yourself in the house, and found some common ground. They like you, they seem to trust you, and they feel good about your company. They have answered every one of the three hundred and twelve questions that you asked over the course of the last, I don't know, nine hours. You've eaten lunch and dinner with them, and even helped that kid with his math homework. It's like you've been dating so far, but now it's time to get serious; it's time to propose!

"What should my proposal look like?" you ask. Let me answer that by reminding you that your primary goal (!) should be to solve your customer's problems, *not* to simply sell stuff. If you keep that in mind, you won't go wrong. In your fact finding you uncovered all of the things that were important to them, and your proposal should be organized, or written in a way that can easily connect your solution back to their problem. For example, if they were primarily motivated by energy savings, you would begin by describing how efficient the new equipment is, and highlight the potential energy savings, if possible. If they were mostly concerned with the house having hot/cold areas, then you would begin by laying out your ideas for re-zoning the house.

Whatever they said was most important is what you emphasize. Solve their problem(s); don't just sell stuff; it's obvious when you do.

There are many ways to present or lay out a proposal or an estimate. But before you settle on which way is best for you, let's continue with some more of the basics. The next "basic" on the list is the brand. Many of you out there insist on selling one of the big, "well-known" brands, like "Barrier", or "Plane", or "Rude". But before you try and convince me that yours is the best, and that "that's all that the people around here want to buy", ask them what brand they have now. Go ahead, ask 'em. What did they tell you, a Honeywell? I once heard of a survey that was done for just this purpose, and they came back with some ridiculous number like 78% of the people surveyed said that they had a Honeywell heating system in their home. Uhhh, nope. Honeywell doesn't make heating or cooling systems, you know that.

The point is that there are an awful lot of contractors out there that will fight me to the death because they think that the brand is one of the most important components of the sale. In fact, if you take that away from them, they will not have the confidence to close a deal. I'll bet you think that's not true, but I assure you that it is. But, the survey SAYS… eehhnntt (you try and spell that), BULLSTEIN! The general public doesn't know one brand from the next, and if you can give them a reasonable explanation as to why your brand is good, they'll take it. They trust you, and want to believe in your recommendation. After all, you are the professional, and they are buying YOU first, and then pretty much whatever you suggest.

Yes, yes, you know a guy that will only buy a "Schnellendorf" because his brother told him that's the best and only one to have. I get it, we all have a few of those, but that's 5% of the public, at best. Don't create your selling style by the exception to the rule. When that guy emerges, sell him a Schnellendorf and all of its splendor for the additional $2,000.00, and be done with it.

The next "basic" to speak of is the idea that you never want to present more than three options. Now you can look at this many different ways, as far as what to include or not include, but in general, you want to be able to present a "good, better, best" scenario. That could really mean almost anything from three different brands, to three different efficiencies, to three different option packages with the same basic equipment, and on, and on, and on. Much of this will vary from one appointment to the next, but you should be getting a good idea of what direction that you're going to go during fact finding stage as we said before. But don't give them more than three options at a time, or they'll never be able to decide. Prospects that were ready to sign the deal tonight will suddenly be overwhelmed by too much to think about, and will have to think about it some more.

Another way to do it would be "A la carte", so to speak, and start with them choosing the equipment, then down the list of options and so on. This works in some situations, but can be time consuming and cumbersome. You'll find that most people don't want to be that involved. They don't care about this stuff like you do, and they want you to make it simple for them. It's ok to let them know that they can change things around within the options that you recommended, but most won't. They trust you to make their problems go away.

Ah, yes, let's chew on that piece of fat for a bit. What exactly do they expect from your recommendations? You will be well served to listen closely here, and avoid many future problems. We have established that you are the professional, and they're not, right. Well then, I strongly urge you to specify what your system will do, and the conditions in which it will do it. This will seem unnecessary to some of you until you get the call from someone that says "It's 105 degrees outside, and my house won't get any cooler than 78! Get over here and fix it now!" I've had that call, and I don't want it again. Hopefully you already know where I'm going with this, and if you don't, you need to learn it, like now.

One of the great misconceptions out there is that air conditioning systems are unlimited. Not that people actually think that it can do anything, but they expect it will always be able to cool their house. But will it? If you were to talk to the engineers that design the equipment that you sell, they will tell you that if the system is designed properly, it will be able to cool the house to, let's say 72 degrees when it's 94 outside. And for this area, that covers about 99% of the cooling season.

But, when it's more than 94 degrees outside, the temperature inside may begin to climb. Granted, that's less than 1% of the season, but when it's happening, nobody cares about your 1%.

You not only need to learn about design temperatures, and how to explain them to the homeowners, you need to learn how to put it in writing on the proposal. If you don't, you may end up with some cranky customers and a lawsuit, or worse. By the way, the answer is not to just increase the size of the unit, or you'll risk under-dehumidification, and shortened equipment life. This isn't a technical or sizing class, true, but you want to do the right job for your client, so be sure that they understand what they're getting, and that you can point to it when they forget.

That was one reason to be very specific on your proposals. There is also a reason to be vague in parts of it. I'm not trying to be sneaky here, or get away with anything, but there are times when you are not sure how you will end up doing a particular task until you get in there and do it. You may think that you will need to cut a certain hole, or use a special tool, but when you start, you find a better way; installers go through this every day. If you say that you will do something on your *legal contract* (proposal), and you don't do it, you may be accused of trying to pull a fast one, and be expected to give them a discount for not doing it.

Many of your competitors will go to look at a job, run out to the truck, and five minutes later come back with something like "One new 100,000 BTU gas furnace - $3,000." The purpose of this book is to turn that mess into a $10,000 sale, with repeat business, and lots of referrals and happy customers; and a much improved closing ratio. But, truth be told, it takes some work to get better and to do it right. That includes spending some time crafting a complete proposal, which should also include many other items that we haven't touched on yet, such as the work area. Which area(s) are *not* included in this project? Be clear about the room over the garage, the back porch, or any other areas that may not have been clear.

Details of the equipment and options should be noted here, as well as any possible add-ons that you discussed, but they weren't ready to buy. These are things that they could add on before the job is complete and perhaps later on for a little more money. How about sub contractors? It's not enough that she said that she would have her brother take care of the electrical tie-in, put it in the proposal;"all line voltage electrical work and materials by others". What about work by others that isn't finished by when they said it would? If you have to make another trip back to do a start up because the electrician didn't finish on time, who will pay for that? Put it on paper. Any other assumptions should be mentioned, too, like if you need to cut a hole in that partition wall to run a duct, but it turns out to be 18" of concrete; who pays for the extra labor? This kind of stuff comes up all the time, especially if you're working in older homes. You'll never be able to cover all of the potential problems, but you need to try and catch the big ones.

You'll want to add in any warranty or service contract information, along with any other special or customary items you spoke of. Then, naturally you need the price and how they agreed to pay for the work. This is one place that you want to be very clear on, such as the down payment, progress payments (how much and when they are due), and final payment.

On larger jobs these payments can become subjective if you are not careful, and can put you in a world of financial inconvenience or worse. Take, for example, the final payment; when is it due? This may sound silly, but imagine your crews have worked on this $30,000 job for two weeks, they have all systems running, are packed up and cleaned up, and now ask for the final $15,000 check. The customer says that he will pay you when you are done. "But, we are done", you argue. "No you're not. I still don't have the custom 2"X8" blue grill you ordered for under the cabinet in my closet" he says defiantly. "But you know that won't be in for another four months!" you cry. "Well, when the job's done, you'll get all of your money." Now what? Avoid this by specifying a small amount to be held back until a punch list, agreed upon by you and the homeowner, is completed; maybe $500.

We're not done here yet; let's go back to the main part of the proposal. When presenting an estimate, you'll want to include everything that was discussed during the fact finding that was important to them. Every option, every box that they said was imperative. Then put the price on the bottom, or the next page. One of two things will happen. First, they may say "OK", and you will fall flat on your face after hitting your head on their coffee table, in complete and utter astonishment that they actually went for it all, or, they say that they can't spend that much, and you proceed to remove things one by one. (We'll discuss the closing techniques later). The point is that if you propose it all, they might say "yes". You can always remove things later, if needed. If you start with the price of just the furnace, for example, it will be difficult to up-sell, because in the split second that they see the price, *that* becomes the number for them to deal with; anything more may be perceived as "not needed".

The rest of this goes back to being professional. If you can do all of this on your laptop, and print it out while you're at the house, you're golden, and on your way to the close. If you can't figure it all out on the spot, set your next appointment NOW! This is not an option, burn this into your brain, don't leave until your return date and time is set, and make sure that all decision makers will be there, otherwise you're wasting your time. When you do return, impress them with your presentation again. Be consistent. I knew of one company that literally had leather bound presentation books that told the company's story, and they topped it off with similar (although not leather bound) proposals. They were so professional and better than the competition, you almost *had* to buy from them.

Last but not least, don't be intimidated by the price that you give them. If you flinch here, you're dead. Don't act like it's a lot of money, or make apologies for all of the zeros. This is what they asked for, how do you know what they can afford? Just because you can't rub two nickels together doesn't mean they don't have ten million dollars in the bank. You don't know; don't think you do. Seriously, if they can't afford it, they'll tell you. If you come across as intimidated by it, they'll think it's too much for you to handle.

Take-aways
- Solve their problems, don't sell stuff
- Prioritize your proposal to match their wants and needs
- Start with everything, you can always remove some
- Never present more than three options
- Don't skimp on the custom proposal materials
- Be specific on some things, and vague on others as needed
- Despite what you believe, brand is not that important
- Try to anticipate unknown obstacles
- Be clear about what is not included
- Discuss the consequences of others' non-performance
- Be very clear on how and when payment is due
- Price always goes on the bottom, or next page

How to Pay for it

What is the most stressful part of any major purchase? Come on, now, this is an easy one. Think about any large purchase you've ever made that involved a salesman. It's paying for it! The interesting part is that for many people, the price isn't the scary part, that's just a number. The scary part is figuring out how they are going to pay for it. Let that sink in for a moment.

The price isn't as much of a concern as how they are going to pay for it. If you're conscious, it may strike some of you that you've been barking up the wrong tree all of these years. You've been focusing all of your efforts on reducing the price when you can't close a deal, rather than helping them pay for it. When most of America goes out to buy a new car, they are looking at the monthly payment way more than the actual price. Who do you know that is writing a $30,000 check for a new Subaru? The answer may not be "nobody", but there aren't that many to be sure.

This is a significant difference in the way many of you currently look at sales, and is worthy of some introspection. I am saying that in many cases you are reducing your price, and hence, gross profit, for no reason! Because... the price is not the problem. Of course, the estimate is more than they were expecting; if it's not, you probably left some on the table, but the real issue at hand is that they didn't want to max out their credit card(s), or something along those lines, and they don't know what to do. They don't know how to solve the problem.

Aren't you supposed to be solving their problems? Well, then why do so many salesmen drop the ball here, and say "when you figure it out, give me a call". I've got news for you; the company that can solve this problem and reduce their stress is going to get the job, even if it costs an extra $500, or $1,000.

Make no mistake, this is the most important part of the job for you. Yes, we want to do the right job for the customer, and solve their IAQ issues, but if at the end of the day you don't get paid, there's nothing else for us to talk about. You're out of business. A big part of a salesman's job is to secure payment, and when working with homeowners, this can be downright paralyzing for some. You need to be able to offer solutions quickly and confidently to secure the deal.

It would be great if all of your customers just wrote you a check at the end of the job, but only some will. You have to make it as easy as possible for the public to buy from you, so you will do yourself and your company a tremendous service by spending a little preparation time here, and most of it you only have to do once.

First off, anyone that doesn't accept credit cards these days is living in a cave. Yes, we would all like to be a "cash only" business, but that's not practical. By the way, how did you pay for this book? Credit card, most likely, but what about PayPal, or other forms of online payments; have you looked into them? Perhaps your company offers some form of in-house financing. Even partial financing could be a huge help to sealing a deal.

Without a doubt, the greatest help to your business, though, will be when you can offer your customers financing through a third party. Many banks and finance companies have programs to finance this type of work, and are quick to set up and get started. The application process is done on the phone in the customer's house, and usually takes less than ten minutes. There is little paperwork to be done, and when the work is complete, your company usually gets paid within forty eight hours, direct deposited to their bank. In addition to the simplicity of it all, these companies usually offer different programs for you to offer your customers, such as no payments for three months, or no interest for XX months, and so on.

How to Pay for it

As good as all of that financing sounds to get your job paid for, some of these companies will do even better than that by giving your customer a line of revolving credit, including a credit card with your company's name on it. This card can be used to finance future purchases or even emergency service calls, *only from your company!* So, not only did you secure your sale, but now who do you think that customer is going to call the next time they have a service issue if they have a line of credit with your company? No, I'm not tellin'; guess.

If you happen to be doing commercial work, it becomes a little more difficult for you to assist in financing, but not impossible. I have yet to come across more than a couple of contractors that offer any financing at all for commercial customers, in fact most don't even know that they can. The way to do it is through a commercial leasing company. These companies will finance almost any type of equipment and its installation. Search the web for a few of these, and tell them what you do and that you would like to offer their financing to your customers. These guys are not used to contractors trying to bring business to them, so they will jump as high as you say, usually.

These commercial leases are figured out on a one by one basis, of course. They will design a lease for a certain period of time, say ten years, and then usually have a one dollar buy-out at the end. Not only will this process alone land you some jobs because you were the only company able to do this, but this will usually create some advantageous tax deductions for your client, and once again, the competitive price issue now becomes diminished because you can sell a monthly payment, when everyone else is selling a total cost. These two paragraphs alone are worth a hundred times the cost of this book. I accept tips.

Let's talk about discounts. I'll bet most of you use them all the time, which is fine, but I want to give you a few things to think about first; things like net profit. Do you have any idea what kind of net profit your company sees on each job? Not gross profit, which is basically the price of the job, minus the cost of labor and material. I'm talking about net profit, which is what I just said, minus all of the overhead like the cost of the trucks, insurance, advertising, etc, etc.

Most companies of our type shoot for a net profit of 10%, whether they know it or not. However studies show that the average is actually closer to 2%! That's true; so, on a $1,000 job, the overall average net profit is around $20; that's it! I realize that most of you don't believe that, but ask the owner. If your company is large enough to track those numbers (everyone should be doing this), he'll tell you.

For now, let's assume (don't say it) that your company has a net profit of 10%. Times are a little slow right now, and you've got a bunch of proposals out there that aren't budging. Your sales manager comes to you and says "we need some work, see what you can stir up". You pull out the $5,000 estimate for a new boiler (not a water heater) that you gave to Mrs. Jones last week and tell her that if she can do it Thursday, you'll take $500 off. She says yes, and your guys can go back to work.

Your discount was literally all of the net profit on that job; and that's because you guys are really good, and have a 10% net. Everyone else that has an average 2% net profit literally lost money on their overhead. Now, I'm not saying that you shouldn't have done it, it might have been the lesser of two evils, but you need to understand the numbers, if you're going to be good at this sales stuff.

Discounts are an important tool, but you need to understand that any discount comes directly out of the *net* profit of that job, as I said above. If you do give someone a discount, make sure that you either get something in return like we did from Mrs. Jones, or have a reason for the discount, and can explain it. What's that? What did we get from Mrs. Jones, you ask? Well, we didn't just give her a discount; we offered her $500 off *if* we could do the work on Thursday, and only Thursday. We gave her something in exchange for filling a gap in our schedule which might have caused us a greater loss for the day, and sent two men home.

If you don't receive something, or take something away when you offer a discount, you must have some other reason for lowering the price, otherwise some will accuse you of charging them too much before, or ripping them off. I know, stores don't have to explain themselves when they have a sale, but in the contracting world you will get asked about this, and if you don't have the right answer, they will scratch you off of their list.

OK, what else do you need to know on this subject, hmmm? Oh, as I said earlier, you need to specify in the contract all of the payments that you will receive for this work and when. (By the way, a proposal becomes a contract when the customer agrees and signs it). This will normally include a down payment. Now, make sure that you know the local laws governing down payments, some may limit you to ten percent, but they will probably allow for an additional payment upon delivery of the equipment and material. Down payments and progress payments are one area that you really need to stand your ground.

This can become difficult when everything is planned out, and material and workers are sent to the job, and then the customer doesn't have the check that they were supposed to have. You will probably want to start the job anyway, because the customer said he had an emergency, but he'll have it for you tomorrow. Red flags and sirens should be going off in your head like mad. You handle this however you think you should, but after a lot of experience, I would tell you to walk away and come back when the check is ready. If you start the work, and they don't have the check tomorrow, then you don't want to stop and get them angry because then you may never get paid, or get back in. That's a bad and sick feeling, if you've ever been through that.

One time, like a hundred years ago, when I attended a sales training seminar, we were doing some role playing on closing techniques in front of the class with the instructor as the customer. Part of the training was a belief that you should stick to your price, and not succumb to pressure for a discount. This was the price, and if you want it to be less, we would have to remove something from the contract. I stood my ground and thought that I had beaten him; then he asked me if we took American Express. I said yes, again anticipating my success, when he said, "ok, I'll pay cash, and you can reduce your price by the four percent that you would have to pay AmEx". I still don't know how to overcome that one.

One last point on payment, before we wrap up this chapter. There may be rebates available from a manufacturer, utility company, or even a government agency that could help offset the price for your customer. These shouldn't be too difficult to find, but you'll want to be able to offer these at the customer's table at the time of the proposal to help you close the deal, so figure out a way to research these on a regular basis so as not to miss any. The key to it all is to make paying for it as easy as possible.

Take-aways
- Figuring out how to pay for it is stressful for your customer
- They're less concerned with the price than the payment
- Offer many ways to pay, especially financing
- Get involved with a company that gives your customer a revolving credit line for use at only your company
- Find a commercial leasing company to work with
- Understand where discounts come from
- Have a reason for a discount, if you offer it
- Be militant about payment schedules
- Don't forget the rebates

Closing Time

Now it's time to show us what you've got; what you're made of. Will you be able to seal the deal on your terms, or will you beg and grovel for any morsel that they will spare for your pitiful self. Well, what's it going to be, boy? Are you going to feast or starve?

That seemed a little harsh, so I'll dial it back a bit. Closing the deal is where the rubber meets the road; where you find out how good of a job you've done up to this point. It's almost like getting graded in school. If you've followed the plan, solved all of the problems that you could find during your fact finding mission, and figured out a reasonable way for them to pay for it, you will have a very good chance at closing your sale.

If there is one chapter that people would turn to first, instead of reading this cover to cover, it would be this one. Everyone wants to find out if this is the book that will change their life with the emergence of the elusive, yet magic closing technique. As if, with these spell induced collection of syllables you will place your prospects into an infallible closing trance in which a signature is certain, and unicorns fly by and deliver pepperoni pizzas to all, amen.

Uh, I hate to be the wet blanket here, but I've never even heard of that happening in a fairy tale, never mind in the HVAC world. I'm sorry, there is no magic to be had. BUT, I'll give you the closest thing to magic that you will ever hear in the world of sales, and this *is* true; **ask for the sale**. All funny stuff aside, the most important change, and I mean *change*, that you can make in your career is to ask for the sale. Give me a minute to explain myself.

I have worked in many different positions in this industry for somewhere in the neighborhood of twenty five years now, with about half of that as a sales manager. I've managed salesmen on the contractor side, and on the wholesale, supply house side. I've even managed salespeople in other industries, and the consistent mistake I see made every single day, is not asking for the sale. I'm not kidding. You want some magic words, try these: "Would you like to buy this now?" Sounds ridiculous, right? Well, welcome to the undisciplined world of sales.

After sitting through thousands of sales calls, what I have observed is that most salesmen never ask for the sale. Those that do, only ask once. Did you know that most deals are closed after the *fifth* attempt? That's right; most deals are signed after five attempts to get the prospect to say yes. These people are all looking for "better" words to help them close, when they're not using any! Many of our colleagues spend hours on a call, and then hope that the homeowner says "wait, come-back. I want to give you my money" as they chase the timid salesman into the driveway.

You may have heard of the "ABC" technique, also known as "Always Be Closing". Not really a technique, but it's better than the more commonly used "ANTCY" technique; that is "Ain't Never Tried Closin' Yet", championed by most. Or maybe you've experienced the "IHNBIWMP" otherwise known as "I Have to go Now Because I Wet My Pants" method. WTF!!! Grab life by the pen and get a damn signature already!!! Ask for the f...... sale!!!!!

While there are no actual secret words or phrases to ensure a signature, there are definitely things you can learn and do to improve your closing ratio. If, for instance, you can get good at all of the stuff in the previous chapters, your close is already half done. However, there are some universal tips that most experts agree will help.

First of all, you want to get them in the habit of saying "yes" as you're approaching your final questions. Think about your questions before you ask them. How are they going to answer? If you're not sure, reword them to get an affirmative. If you ask the homeowner "So, do you think that you want to go for the extra $1,500 for the 16 SEER unit?" you can't be sure what his answer will be, but there is a good chance that it will be "no". Instead, if you were to say something like "Mr. Vinci you said that you were interested in the energy savings of the 16 SEER unit, isn't that right?" you are pretty sure that you'll hear a "yes", because he already told you that he was. Subtle changes can have a huge effect on the outcome.

Next, watch their body language. Somewhere around seventy five percent of all our communication in life is non-verbal. That means that most of what you learn from people doesn't come from what they say, right? So that means that you almost certainly need to do a better job deciphering their body language. If they have their arms crossed, and are leaning away or back from you or the table, they are not accepting, or open to what you are saying. They may disagree completely, or just aren't ready to decide yet, but something is wrong, they are showing you a defensive posture, and that will not produce a signature.

Naturally, the opposite is true as well; if the husband has his arm around his wife, or they are leaning forward, intently listening to what you are saying, that's a universal buying sign. That doesn't mean that you'll definitely close the sale, but they are willing to continue the discussion, and will allow you to try and further convince them. The important thing here is for you to refrain from trying to close until you see positive body language. If you try to push to close too soon, you may set them on the path of saying "no" and make them defensive, which, once in motion, is very difficult to reverse.

One of the things that I like to talk about, when teaching closing techniques, is the idea of asking "A or B" questions, rather than "Yes or No" questions. This is simple enough, just think about what you're asking. If you ask them a question like "Are you ready to start the project on Monday?" you are giving them an easy way out, or a chance to say "No, we're not ready yet". A better way to ask that question would be something like "Is this Monday a good day to get started, or would you rather wait until Wednesday?" This could also be classified as the "Assumed consent" close. This technique has the salesman assume that what they have discussed so far has been agreeable, and that the customer would have let you know by now if they had any objections. In other words, the customer has already agreed to the project non-verbally, so now you just need to handle the formalities.

Don't be concerned about stepping over any boundaries here. As always, be polite and agreeable, and certainly not pushy. If you have read their signals wrong, and they don't want to proceed, they will tell you.

Another closing technique is to ask if there is anything that they are unsure about or have objections to. Here you are actually asking for them to voice any obstacle that would keep you from going forward here. If they don't have any, then you can *assume* that they want to move forward, and find out how they would like to pay by asking "Do you want to give me a check for the deposit, or would you rather finance the entire project?

By the way, never say something like "If you can't afford to pay for it, we can ..." If it isn't obvious to you already, pay close attention here. Using the words "if you can't afford it" is insulting to most folks, and even if it is true, you have probably blown the sale. You must learn to be diplomatic with sensitive subjects like money. It doesn't matter if you don't understand why, just understand that it is.

As we have established, there simply is no magic here, that is, other than to know that you must actually ask for the sale. It doesn't mean you have to say the words "Will you buy this now, please sir?" But you do have to ask some type of closing question. And, know that you probably have to ask numerous times before they finally commit. This would likely be a big purchase for anyone, so it's only natural for them to need a little time to decide.

Let's go back to the statement I made before about having to close five times on average. Some of you were uncomfortable with that because you think that I meant that a salesman said "Are you ready to sign the contract yet?" and the prospect said "no". Then the salesman asked again and again, each time with the homeowner getting progressively more annoyed and raising his voice "NO, I SAID!!!" until, out of energy, the customer finally gave in, exhausted, said alright. That's just what a good salesman has to do to be successful.

No, that's not at all what I meant. What I meant was that a salesman would ask a closing question like "should we put you down for the electronic filter, or do you want to see if the media filter would be sufficient for Mary's allergies?" and rather than answer the question, which would assume that they were buying the system, the customer redirects the conversation. Usually, it will be to ask another question, and ignore the salesman's implication all together. It is important to note that the customer doesn't shut him down by saying that he won't buy, he just isn't quite ready to say "yes" yet.

That counts as an attempt to close. So the conversation will continue, and when the salesman thinks that the prospect has the information that he needs, he'll try to close again by saying something like "Have you decided where you would like us to put the condenser?" This is very subtle because he is implying that "we" will be placing the condenser. If the homeowner says that he would like us to put it on the south side, another closing question should follow quickly to confirm his intent. "Sure, we can do that. What time would you like us to be here in the morning?" If he had said "I don't know where I want the condenser yet" and changed the subject, the dance would need to continue.

There are other subtleties that play a subconscious role in assisting in closing, or not closing a sale. Saying things like "Hmmm, I never did one like this before" or "Wow, that's going to be difficult" are *not* helping you at all. Talk like that gives people the idea that this may be too much for your company, and that maybe they should call someone else. People want to buy from someone that brings confidence and security to the table, not uncertainty and nervousness. Perception truly is reality, and if someone thinks you're in over your head, then they'll get someone else. Ironically, the other guy will probably not have any more experience with that type of work than you; he'll just choose the right words.

Along the same lines, using a phrase like "sign the contract" while accurate and truthful, may be intimidating to some. You can get the same result by saying "we just need to do a little paperwork", and explain all of the legal aspects needed. It's the same thing, just with a lighter tone. The point is, some people want the facts shot at them right between their eyes, while others need to be finessed and massaged, and you will meet them both.

Remember, the prospects called you because they wanted to get this work done. They know that they want to do this. If the close isn't coming easily, then you probably missed something earlier in the process. But really, there are only a few possibilities. You either didn't connect with them personally, you didn't offer a satisfactory solution to their problem(s), there is still a money concern, or they are not confident that you can do the work. Don't be afraid to ask them directly, then circle back, and go for the close again.

Take-aways
- Most salesmen *NEVER* ask for the sale
- There is no magic technique; ask for the sale
- Learn a few simple techniques like assumed consent; ask for the sale
- Always "A or B", never "Yes or No" questions when closing; ask for the sale
- Most sales require five attempts to close the deal; ask for the sale

- If the close isn't imminent, you probably didn't adequately solve one of their problems; go back and review
- Ask for the sale
- Ask for the sale
- Ask for the sale
- Buy more books for your friends
- Ask for the sale
- Ask for the sale

Overcoming Objections

One of the most common topics of training and discussion in the sales world is that of overcoming objections, second only to closing techniques. In order to learn how to best respond to these impromptu hurdles placed in our passageway to a signature, we must first understand what they are and where they come from.

An objection by a potential customer in the sales process is a person saying "You haven't given me enough information, solved my problem, or convinced me that you can do the job or that I can trust you, or I don't know how to pay for it". That's pretty much it; anything else that you can think of will fall into one of these categories. Of course, there are always personal preferences and quirks that no amount of counteraction will defeat. If the customers says that they will only do business with people born on February 29th, well then, just leave and move on.

Let's begin by assuming that we are only dealing with rational people now (Ha, good luck!), and discuss what objections that these "sanies" might have. I know, they shouldn't have any at this point, since you did such a fantastic job in your presentation and fact finding, but let's just make believe that their hearing aid was on the fritz, and they missed some of your magic. How will you, or should you handle them? Well, for one thing, you need to treat each and every concern that they have as valid, and never dismiss them as anything less.

The most common *voiced* concern is usually that of price. I say it that way because they will sometimes use price as an excuse, but that is often just a "catch-all", and it really could be almost anything. If it is actually a money or price issue, you now need to determine if it is just a matter of how to pay for it, or is it a value problem, and they are unclear that what you proposed to them is of equal or greater value than the price you quoted.

If it turns out that your customers are concerned about the actual amount of money that their desired system will remove from their bank account, you are in luck. That is a good problem to have. What they are telling you is that you did a good job with your presentation and recommendations, and they want it all; now it's just a matter of what combination of cash, credit cards and loans or financing will you use. It may take a little bit of effort on both of your parts, but the deal is done; you have a solid signature. The worst case scenario here is that maybe you only do part of the work now and the rest in a few months.

The other, more common money/price objection isn't actually so much about the money, as it is about the value. This objection is telling you that you either didn't convince your client that your recommendations would solve their problems, or you missed on the fact finding part, and didn't properly understand the problems as they see them. Regardless of the exact issue, you should go back and confirm and review the problem that they want to solve, and then give a better, perhaps more detailed, explanation of why you prescribed what you did, and how it will solve the problem.

This could even be an issue with you offering a more expensive brand than they wanted causing all of the fuss. That still qualifies as your solution not fitting the problem. Now, the "problem" in this case may be less of a problem per se, and more of a preference (that being an expensive brand), but either way, you misunderstood their problem (preference) for a less expensive brand of equipment during your fact finding process.

And while we're on this subject, let's take a minute to remember that there is an ass for every saddle, excuse me, a hiney for every saddle; do you agree? It takes all kinds of buyers out there to make the world go around, and you will sell to most of them in your career. Buyers that only want the best, some that only want the cheapest, some that don't care as long as it's blue, and so on. Once you're in the house, you can give a gentle opinion, but you should present the options that best suit their wants and needs, and let them decide for themselves. If they want an inexpensive unit, and you only offer them the big, well known, national names, then you should expect a price objection. Don't impose your preferences on their budget.

How about the couple that says that you're $2,000 more than the last contractor that was in here, how do you handle that? Well, assuming that you are not typically charging double what everyone else does (I'm not saying you can't or shouldn't), It's quite likely that the other company didn't go through the same comprehensive fact finding process and discover all of the issues that the homeowner would like to improve on, and is just quoting a simple equipment swap out. Of course they're $2,000 less; they're only doing half of the work. Do you want a contractor that would sell you a new microwave oven, only to find out that the last one didn't work because the power was turned off? I doubt it; that guy will give you a new piece of equipment that runs like new, but still gives you the same hot and cold areas of the house.

Interesting point here; most homeowners don't have the slightest clue of how their heating and cooling systems work. They may very well actually think that by simply installing a new furnace, all of the hot and cold spot problems in the house caused by ductwork issues will automatically be resolved. I would be sure to make them understand that is not the case, and any good HVAC contractor would let them know that.

One of the most difficult parts of being a salesman is when you can't actually get to the objection. Less experienced salesmen are often intimidated by closing objections, when in fact the veterans welcome them. Why? Because if the client gives you an objection, you have the opportunity to overcome it. If Mrs. Stromboli says that she doesn't want to buy from you "because you said earlier that your company makes every new employee take drugs", and she doesn't like that. "Uh, no, I said we make them all take drug *tests!*" "Ohhh, ok; that's better. Is it ok if I pay you in cash?"

But if you can't get the client to tell you what they're uncomfortable with, you're stuck. If they allow you the time to go back through it all, you might have a chance to figure it out, but that's a long shot. If they just won't give you a clue, then chances are good that they already found someone else to do the work, or they just didn't like you or something about you. Don't take it personally; it happens to all of us. If it happens more than once every twenty five or so calls though, you need to re-examine your mannerisms, or take a shower, or something.

As you have probably already heard through the grapevine, by far, the best way to overcome any objection is to address it before they do. As you go back and look at your successes and defeats, you will start to realize that when you were able to do your complete presentation, with the book of photographs of your work, and the newspaper articles, and thank you letters to your company, along with a full and detailed fact finding mission, and the proper solutions and recommendations to their problems, the objections were few, and the dollar amount was high. You'll spend more time filling out finance forms than answering objections.

One of the objections you never want to hear is "Ok, that sounds great. I'll go over it with my wife when she gets home, and we'll be in touch". You just wasted your evening. Never go on an appointment without all of the decision makers present. Even if the one was prepared to say yes to the furnace swap out, they will never agree to all of the extras without their partner, and worse, they will absolutely never be able to convey to their spouse what you got them all excited about. The only thing that they will remember is that you were a nice guy, but you were $1,800 more than everyone else, and nothing about the $80 a month in fuel and electric savings. So, no "one-leggers", got it?

Take-aways
- Objections are good; answer them well, and you've got a deal
- Unless they sign right away, a lack of objections is bad. You don't know what the problem is

- Most objections arise because they don't know how to pay for it, they don't see the value, you didn't solve their problem(s), or they don't like or trust you
- If you're *way* more than the other guy, they're not comparing apples to apples; tell them!
- An appointment without all decision makers there is a waste of time

ADD-ONs

Add-ons to a sale are one of the least understood, yet arguably one of the most important pieces to the sales puzzle. If more owners, salesmen, and technicians took the time to analyze and understand the financials of what they do for a living, they would all make a lot more money. But as usual, they're all too busy to make money; they have work to do! Again, why do you go to work every... never mind.

Personally, I am a big fan of doing *less* work, and making more money, and I'd like to attempt to, gently, bash that into your head here. A few chapters ago, I said something about service managers being more concerned with technicians hurrying through each call because they have so many on the board on those really hot or cold days, that they don't take the time to actually make the money that they should. Well, equipment installations can fall into a similar category of financial malpractice.

This is so prevalent in our industry because the employee is just happy to get a $X,XXX sale, that they ignore the four or five other items that would naturally accompany it. They think that they are making good money on this sale; the optional stuff doesn't amount to much anyway, so why annoy the customer any more than they need to. Hmmm, watch me make them look stupid, now.

To demonstrate just why this is so critically important, I'm going to entertain you with a little math here. Don't get nervous, I'll do all of the hard stuff, you just do what you always did in math class. Err, better yet, pay attention this time.

Ok, here we go. Let's say you sold a gas furnace replacement; nothing crazy here, just a nice simple swap-out, for say, $3,500. I know that already there is someone out there saying "phfffph, that's nuts! We would never sell one for less than $3,600!" Just sit down and play along there, Sparky. You can put in your numbers later.

As I was saying, you sold this furnace for $3,500. I want you to see how the numbers actually break down on a sale like this for the average company today. In rough numbers, all of the material will cost around thirteen hundred dollars, the labor, around six hundred fifty, and then there is the overhead. This is where everyone goes wonky. Somewhere, somehow, someone has to pay for the phones and the desks, the paper, the people in the office, the building, the trucks, the uniforms, the paper plates, the training, the inventory that you dropped in the puddle, the advertising, the printing, the salesmen, the warehouse guy, the incentives and spiffs, the gas bill, the water and sewer bill, the pens and staples, the computers and printers, the internet, the web page, the owner's salary, the accountant, the property tax, the income tax, the licenses, the electric bill, the truck repairs, the radios or cell phones, the dispatcher, the answering service, the alarm company, the lawyer, the shelving, the tools, the machines in the shop, the ladders, the forklift,…get the point? The overhead.

Overhead can be figured different ways, but for the purpose of this exercise, we're going to set it at $1,200 per day, for a two man crew, which seems to be close to average. So, now we know the cost of the material ($1,300), the cost of the labor ($650), and the overhead ($1,200) for this job. Do the math real quick, and you will be left with the net profit; a measly $350, if nothing goes wrong. As soon as you have a call back because someone forgot to take a piece of Styrofoam out of the blower wheel, you're down to a net profit of $86.42. Hardly the "thousands" that everyone thinks the company is making.

Sale price - $3500
- Material cost - $1,300
- Labor cost - $650
- Overhead - $1,200

Net profit - $350

So, now I want you to see what happens when we add on something simple and easy to sell, like a basic media filter. Using the same $3,500 furnace sale, let's say that this time you focus on the virtues of cleaning the air in the home by adding a media filter on to the sale. The homeowner likes the idea of helping reduce the effects of their daughter's allergies, and also that they may not even have to dust quite so often. They agree, the extra $450 is well worth it.

What happens next? You add the media filter on to the material list and have the installers, well, install it. Now let's examine how this will affect your net profit. Let's see, the material cost will be $100 if you have a really bad buyer in your company. Don't tell me that there is *any* additional labor cost here. You know as sure as there is hair on a bear's butt that putting a media filter on to that furnace will not cause those two techs to stay one second longer than they were going to in the first place. They got there at nine o'clock, and they'll leave at three thirty if they have five furnaces to install; don't kid yourself.

The only thing left is the overhead. How much should we account for here? Zeerroo. That's right, zero. Remember, we said that we would figure a cost of twelve hundred dollars per day, per two man crew. Well, we already have that in the job, there's no reason to put it in there again. So, now let's do the math again.

Sale price - $3,950 ($3,500 + $450)
- Material cost - $1,400 ($1,300 + $100)
- Labor cost - $650 ($650 + $0)
- Overhead - $1,200 ($1,200 + $0)

Net profit - $700

Isn't that interesting? By simply adding on a basic media filter, you have now literally **doubled** the net profit on the job! **Doubled!** I always fear that many guys don't grasp the importance of what they just witnessed. This concept is potentially life changing! If you're a one man shop, and you were to do this every day, you just doubled your annual income! If you made $60,000 last year, now you could be at $120,000! Don't believe me? Go back and look at the math again. If you have six crews doing this every day, I'll expect an invitation on to your new thirty five foot fishing boat in the Caribbean this spring.

As you probably know, car dealerships don't make a whole lot of profit on selling a basic car; they make most of their money on the options, extra warranties, etc. Likewise, you should consider selling a basic furnace as a way to keep everyone working and a way to cover the overhead. The add-ons should be where you make all of the gravy. And, as I'm sure that you would agree, it's much easier to add on a humidifier than it is to find someone to buy another furnace. Add-ons are the difference between eating and feasting.

Hopefully now you have a new found thirst for adding on to all of your sales. In order to do that though, you're going to need to create a long list of optional items to help you choose from so that you can offer three or four in every home. Some will be health or energy related, and others may just be luxuries. Don't prejudge anyone; you just need to present the opportunity, and let them decide if they want it or not.

Your list is probably a little slow in getting started, right? I'll list a few here, and you see if you can come up with any others. Don't forget that these vary depending on which part of the country you live in. Just give it some thought, and try to think outside of the box.

- Media or electronic filter
- Humidifier
- Ultra violet light
- Higher SEER
- Change A/C to a Heat Pump
- Add air conditioning coil to a furnace (also creates a lead for another sale)

- Any work for a sub contractor that you can mark up (electrical upgrade)
- Additional ductwork
- Add a zone
- Replace leaking valves
- Add shut off valves
- Add additional outside faucet
- Add a Hot and Cold hose faucet to the driveway for washing car
- Outdoor reset for boilers
- Communicating thermostats
- New diffusers
- Flood alarm near water heater
- Duct insulation
- Outdoor combustion air kit
- Remote oil tank gauge
- Clean and tune any other system in the home
- Attic exhaust fan
- Pull-down attic stairs

Of course, not all of these will apply to your situation. Others, meanwhile, can and should, turn out to be another job entirely. If nothing else, these will give you a place to start. Get a list of your own going, and you'll be surprised at not only how many you can come up with, but also how many of these things you end up selling once people find out that you offer them.

Next, consider creating different "bundles" or "packages". Consumers love bundles, like "The Professional Indoor Air Quality Package" for $1,995. This package could include an electronic filter, an ultra violet light with odor removal, a powered humidifier, and a high tech thermostat to control it. Or you could do a "High-efficiency oil boiler kit" that might include a combustion air piping kit for the burner, a new programmable aquastat with outdoor reset, and a remote oil tank gauge for $1495. You'll find that they may not be interested in these items by themselves, but make it a "package" and suddenly they have to have it.

You shouldn't feel that you need to do this all yourself either. You can present some of these ideas to your customers at the first meeting, and then arm your installers with a few items to revisit. Very often a homeowner will be interested in your suggestions, but just isn't quite ready to commit to them on the spot. Seeing the item for the second time, though, might do the trick. Even after that you should make notes in a tickler file to remind yourself of other items to revisit with them in the future. Why not give them a call back in a few months when it's slow and try again?

The general public doesn't have a good grasp of, or really even care about your overhead. So, they likely don't understand that it will cost more to do the additional work at another time. Therefore you will be doing your customer a favor by reminding them that it is always more cost effective to do the additional work now, while your workers are already there. You'll certainly be happy to come back anytime for the other work, but you won't be able to offer them the same add-on discount that you can now.

One technique that I see many technicians use to generate add-on revenue is to leave a particular item with the homeowner to revue, while they tend to the original task. For instance, before beginning his work, the tech might see that his elderly customer was having a difficult time reading something, and might ask if they also have a tough time seeing the numbers on the thermostat. Following a positive response, he would run out to the truck and bring back a flyer on large numbered thermostats, and the actual unit to inspect. Obviously, a salesman could do the same.

I know of a UV light manufacturer that produced a great sales presentation on DVD for customers to watch while the salesman did his paperwork or a load calculation. They also did such a fantastic job with placing the entire sales presentation on the packaging for some of their products, that they truly sell themselves. The technician can literally bring one in, hand it to the homeowner and ask them to look at it while they finish up their work. I heard that one service tech used that method on every call and sold over a hundred of them in a six month period!

Another thing we should discuss here is the need to charge for everything that you do. To begin with, there is no reason to give anything away for free, even your time. If you're running some type of sales promotion, that's one thing, but people are realistic and don't expect things for free. They know that material and labor is expensive. So, if the homeowner asks you to "just" fix or repair something while you're there, if it will take more than one minute, tell them that you would be happy to, and that you won't charge them *much*. And you don't need to charge much, but charge *something*, otherwise they may continue to ask, and expect it all gratis.

Yeah, yeah, that's right! And another thing; just because you know how to fix or install something and it happens to come easy to you, doesn't mean it's like that for everyone; that's what having a *skill* is all about. You may say to yourself "this is so simple to install, how can I justify charging them $200?" Well, like I just said, just because it's simple, doesn't mean that they can do it, or even want to, and even if it is that simple, it still costs you time.

Remember, this is how you feed your kids, or your drug problem, or that heard of wild starfish in your bath tub that keep sexually harassing your wife; you're not doing this for fun. Your time costs money, whether they pay for it or you do. Would you pay their landscaper's bill? No, then why would you pay their HVAC repair bill (that's you)? If you want to do them a favor, then come back after work and do it on your time.

Let me get back to the point of the chapter. It's always easy to let one subject morph into another, but these little extras we just talked about count as add-ons as well. The extra profit that you're looking for on each job doesn't have to originate from one of the items on your list. It could be that they ask you to get rid of that extra empty oil tank that they have in the basement, or even put their antenna back on the roof that the wind blew down. It doesn't even have to be HVAC related, it only has to be money related. Make an effort to learn how to add on to each job for more net profit.

Take-aways
- Overhead eats most of your gross profit
- A simple and small add-on can literally double your net profit
- One add-on per day could double your salary
- Everyone should be encouraged to add on to your job
- People love "packages" and "bundles"
- Continue to add on even after the sale
- Don't be afraid of the price; they expect to pay
- Charge for everything, even if it's a small amount

Post Sale Minutia

All hail the king of sales (yes, you, in the mirror!). You did it; you closed a big one, with add-ons and options like you wanted. It started out as a simple furnace swap out, and became a four day project with air conditioning and a zoning system. Good job; but now what; are you done and on to the next one? Uhmmm, nope; not yet.

Now, *most* of your work is complete, but there is still plenty to do and really, you're never completely finished. Think of your new customers as new members of your business family. You want to be the one that they call for *any* problem in their home. You want to be their contact to the contracting world. But, once again, I'm getting ahead of myself.

Once you have a signed contract and an agreed upon plan of action with the homeowners (start date, etc.), the next step is to clearly and accurately convey your vision of this project, and the end result, to the installation team. This involves much more than simply reading the contract to them, and giving them some half-assed drawing you did with a crayon on the back of your rendition of that shelled Leonardo guy, eating a pizza in some sewer. Hey, do you still have those pajamas?

First, you need to generate an accurate material list. I can see that this is making a few of you nervous because you might not have actually been part of a physical installation before. That's ok; if you haven't, just do it with the help of the lead installer so that you can discuss your ideas, and what parts and pieces that you will need to complete the project. After that, make sure that you walk him through the jobsite, usually the morning of the first day on the job, to show and discuss with him every little detail you promised the homeowner, and how it is to be done. One thing that may help him remember all of the details that you're giving him is to bring some three inch wide, blue painter's tape, and place pieces of it in those important spots, with your notes written on it. They are easy to see, and work well as reminders as the workers move from room to room. If you or the homeowner designated a specific location of a ceiling diffuser put a piece of tape there too.

Delivering what you promised, and what is on your contract, is critically important to your success and reputation in this business. That includes the smallest of details, even if you don't think they matter. If you said that you would install 8 X 16 ductwork on the contract, but use 10 X 12 instead because it fit better in the space, you must inform the customer or risk them accusing you of trying to get away with something. Be sensitive to the idea that they don't know anything about our field, which can make them overly defensive and concerned about being taken advantage of. Totally understandable, just give them the courtesy of informing them of any necessary changes before they happen, and why.

I hope you are connecting the dots here, and see that this is one of the reasons that I mentioned earlier about wanting to be purposely vague on parts of the description on the contract. You should use phrases like "properly sized ductwork" instead of a specific dimension. Having to go back later and redo or replace things just because they are slightly different that what you said on the paperwork is the type of mistake that could put your job into the loss category very quickly.

It is in your own best interest to stay involved throughout the entire project and in contact with both the lead installer and the homeowner, often acting as the liaison between the two. You are the one person that has a connection to each side, and has the most to lose if a problem remains unresolved. And problems *will* come up; and often. Most of us are human, and are going to make mistakes. Mike is going to put a hand print on the newly painted wall, that dizzy new helper is going to step through the ceiling again and catch the joist in his groin like usual;

dopey bastard. And, of course the genius at the supply house is going to deliver a bathtub to your a/c job, instead of an air handler; again. What? Yup, third time this week. It's going to happen. Don't get mad, just deal with it and move on.

Face problems head-on, quickly and with confidence and authority. If someone falls through a ceiling (and they will), alert the customer about what happened, and reassure them that you will have a carpenter and painter there to repair it ASAP, that then becomes your #1 priority. Your response to these mishaps can do more for your reputation, in a good way, than you would guess. Prove to them that your company truly stands behind their work, and you will earn a steady stream of referrals.

Nice the way that worked out; from here we move on to referrals. Salesmen are chronically bad at asking for referrals, despite the fact that they are one of your best sources of leads. Not so that you won't need to do any other type of prospecting, that's not what I mean. But when you get a referral that is interested in having work done, your customer will rave to them about how detailed you were, how well you explained all of their options, and how you were always there to take care of any little issue that popped up. And, they love how great their indoor air quality is now, too.

Asking for referrals after they sign the paperwork, as well as when the work is actually completed is not only acceptable, it's integral to your success. One of the easiest ways to broach the subject is to introduce a pre-printed form to them in the beginning of your presentation, explaining that later, if they like what they see from you, you would appreciate a few referrals of friends, family, or neighbors who may be in the market for similar work. Then just leave it there in plain sight. By timing it all this way, they are not pressured into acting right this moment, and it gives them time to actually think of some valid names.

This process is uncomfortable for some in the beginning. Don't sweat it; you'll get better at it, especially when you realize that it actually works! You could even employ some sort of gimmick to make it more interesting and fun, like giving them fifty dollars off their job if they actually introduce you to three of the names on the list, or maybe a free box of donuts or something, you know?

The next box on your checklist of salesman superstardom comes when the work is complete. Someone needs to walk the customer through everything that has taken place in their home and explain what it is, and what they need to do with it, if anything. Once again, do not leave this to the installers if you want any accolades. Yes, they will probably clean up, and collect the check and all of that, but is that how you want them to remember your company or your service? Certainly not, unless you want to be like everyone else, and isn't that the reason you're here in the first place; to be better than the crowd?

Put yourself in the customer's position for a moment. They took a leap of faith to go with your company and all of your suggestions, even though it ended up costing way more than everyone else. They opened up their home to you, and let three strange men take over, and get into every little corner of their personal space without even being able to keep an eye on what they were doing most of the time; a complete leap of faith. Now, the techs pack up their stuff, toss some paperwork on the table as they're running out the door, as one yells "Thank you!" and they're gone. How would you feel? Glad it's finally over, of course, yes, but would you feel good about it all? Hard to say.

That is precisely why you need to be there at the end of the job. If you show up an hour or so before the guys leave, you can walk the homeowner through each part of the house and inspect all of the work. Show them all of the ductwork, point out the dampers, and so on. That way you can both clear up anything that was overlooked while there is still someone there who can fix it. Take your time here and really get them familiar with what they bought. Too many technicians and salespeople assume that the customer will know why there is water coming out of that pipe, or why the red light is on. They don't; tell them.

Next, present them with a customized three ringed binder, or similar, with all of the important paperwork including installation and owners guide, the original brochures that you showed them, warranty stuff, and any and all receipts and service agreements from your company. Put it together, organize it, and present it to them with a title on the front like "The Anderson Family HVAC System - Important Papers". Then review it all with them, including how to work the thermostat, where the filter is and how to change it, what to do if they notice water coming out of the *other* pipe, how often they need to have it serviced, and anything else that you can think of. Now, get the completion signature on the contract, and final payment.

Yes, I know that some of you are thinking that "I don't have to worry about that stuff, Steve's a nice guy, and he's usually pretty good with the customers". That may be true, but he is not the person that the client bought it from, and not the one that they're going to call when something goes wrong. Nor is he the guy who has a vested interest in obtaining a referral and recommendation, or the one who made all the commission; so get out from under the bed, and get your ass back over there!

Lastly, in order to truly develop into a world class salesman, and I'm serious about this, you'll want to put together a program or system, that enables you to keep in contact with them on a regular basis. Optimally, it would be best to do this through multiple formats such as email, actual mail, and even an occasional phone call. You can alert them to sales that your company is doing, mail a postcard at service intervals, send out a survey, or just call to say "hi, I was thinking about you; hope all is well". Send them a Christmas card or a card for whatever alien holiday that they celebrate. (You know all of these things about your customers because you keep a tickler file of all of this information, right?) These are all great times to ask for additional referrals, too. I know that nobody else does any of this, that's the point; that's why they will recommend you to their friends, get it? If you do, you're on your way to great success!

Take aways
- Don't bail out yet, you're not done
- Review the plan accurately with the installers
- Leave notes in areas of importance

- Problems and mistakes will happen; deal with them quickly and decisively.
- Don't trust the final paperwork to anyone else
- Don't leave without getting paid
- Follow up regularly; customers are your business family
- Ask for referrals again
- Remember the starfish

The After*math*

The rest of what I have to say is just a bunch of random thoughts that don't necessarily require individual chapters. However, I think that they are important enough to spend time explaining them, so here we go.

Stop selling on price. If you have to resort to being the cheapest of the five estimates to get a job, you missed the entire point of the book. You bring value to the table that others don't, charge for it. Don't race them to the bottom, in fact, be proud that you're more expensive than those guys. "Absolutely, we're more expensive than *those* guys!"

Look back at your worst customers throughout the years; who are they? I'll bet that they are the customers that insisted on a discount, and business was slow at the time, so you gave it to them. Worst thing that you ever did, right? Put a picture of that guy on your desk, and every time someone asks for a better deal, ask yourself "How many more of these guys do I want?"

The profit is in the details; you can take that to the bank. Remember when we talked about the average net profit being a measly two percent on average? Well, all of your profit is being thrown in the garbage or dropped in a puddle. Think about the sheet metal pieces that get thrown out instead of straightened, or the box of flex that gets tossed with eight feet left in it. How much is 2% on a $10,000 job? $200; hmmph. Step on the thermostat in the truck and now you actually *lost* money on the job. Why do you think that successful old guy keeps picking up wire nuts off the ground? He understands, and the sooner you understand why your net is so low, things will change, but not before. Details.

If your schedule is always full, and you are booked more than two weeks out, you need to raise your prices. Add on five percent to every price; what's the worst that could happen? You make too much money? Get out. Just leave.

How much did the Smith job make, do you know? Why not? If your company is not job-costing every single installation, how do you know if your prices are correct? Your service work could be making all of the profit, and your "high dollar" installation projects actually lose money.

I literally just spoke to a contractor yesterday that told me about how they started doing installations, and how they expected to make a lot of money because they had so much work. When I asked how they were getting all of this work, they told me about how expensive the other guys were, and that they were able to do the work for a lot less money. I asked if they were sure about their expenses, and of course they said "What expenses?" They'll learn.

If you are a salesman, or even an owner for that matter, you need to know what jobs are profitable, which ones aren't, and why. That is the only way you will improve the financial situation for either one of you.

Speaking of commission, are your technicians being paid commission or spiffs? They should be, if you want them to sell. This is where I usually hear two distinct arguments. The first says that they don't want the possibility of guys selling things that people don't need, and that is valid. However, let's not throw the baby out with the bath water, just police it like anything else. Preach honesty in your meetings, punish bad behavior, and things will be ok.

The other argument goes like this. "I pay those guys too much already. They should sell this stuff because it's part of their job". Get rid of that thinking right now, if you want your company to survive. Everyone wants to improve their lives and have an opportunity to do better. Remember capitalism? They're not you, and they're not going to do what you think they should do. (I'm speaking to the owners now.) If you don't want your technicians stealing your customers or doing side work, then you need to pay them extra when they do something extra. By the way, your competition offers a spiff program; how long until your employees find out about it?

Pay your employees well for sales, and you will be pleasantly surprised at the results. This goes for installers and helpers, as well.

<u>Take aways</u>
- Go grab a snack, you deserve it.

What Are You NOT Doing?

We all want the magic bullet, but now you're finally old enough to hear the truth; it doesn't exist. We can go back and forth about Santa Claus, and I can tell you a story or two about the tooth fairy and that little dress she…, uh…, never mind, but the magic bullet is a fairy tale if there ever was one. The fact is, this sales stuff bears a strong resemblance to work, and that's why they pay you for it.

Again I want to congratulate you for taking the time and making an effort to improve your sales skills and career. I hope you aren't too disappointed about the magic bullet thing, but really the resulting benefits can be just as exciting if you follow through with your desire to improve! Start here by asking yourself "what in this book am I not already doing?" The difficult part is always having the discipline to not skip over the parts that you don't like.

Most guys don't want to go back after the sale, for instance, and make sure that the customer is satisfied; that's just an invitation for complaints, you say. Well, yes it is, and that again gives you the opportunity to showcase how good you really are. Things like this make you stand out from the crowd because *no one* likes to do the things that are difficult, and that is exactly why they make you stand out. And yes, people will notice.

So, put in just a little more effort than you are now, and make this a truly great career for yourself. Be honest with yourself about the things that you are missing in your current routine, make the needed changes and you will be happier, well respected, and better paid.

Other Stuff

First, if you would like to reach me, shoot me an email at richardjschuster@gmail.com, or take a look at our website at www.RJSchusterBooks.com. Then, if you get a hankerin' for more, you can check out some of my other books, like *"No Ducks in the Attic"*, or *"101 Ways to Suck as an HVAC Technician"*.

Well, good luck, it was fun, and thanks for your support. Oh, and come to one of my seminars sometime, and I'll buy you lunch.

RJS

www.ingramcontent.com/pod-product-compliance
Lightning Source LLC
Chambersburg PA
CBHW061146180526
45170CB00002B/635